A SURVEY OF
PHYSICAL THEORY

Max Planck

Translated by
R. Jones and D. H. Williams

DOVER PUBLICATIONS, INC.
New York

Published in Canada by General Publishing Company, Ltd., 30 Lesmill Road, Don Mills, Toronto, Ontario.

Published in the United Kingdom by Constable and Company, Ltd., 3 The Lanchesters, 162–164 Fulham Palace Road, London W6 9ER.

Bibliographical Note

This Dover edition, first published in 1993, is an unabridged, slightly corrected republication of the 1960 Dover reprint of the translation originally published under the title *A Survey of Physics* by Methuen & Co., Ltd., London, in 1925.

Library of Congress Cataloging-in-Publication Data

Planck, Max, 1858–1947.
    [Physikalische Rundblicke. English]
    A survey of physical theory / Max Planck ; translated by R. Jones and D. H. Williams.
        p.   cm.
    Translation of: Physikalische Rundblicke.
    Includes bibliographical references.
    ISBN 0-486-67867-9
    1. Physics.   I. Title.
QC71.P72   1993
530—dc20   93–6110
                                                               CIP

Manufactured in the United States of America
Dover Publications, Inc., 31 East 2nd Street, Mineola, N.Y. 11501

Preface

The surveys of physics which I have endeavoured to give on several occasions during recent years, are now collected in chronological order in the following pages.

Since it was my intention that the essays should reach a wider circle of readers in due time, I am now publishing them in an almost unaltered form without much fear of seriously contradicting myself, at least on physical questions. With regard to my occasional references to psychology, my own views have been accepted in certain circles, but they have been severely criticized in others. I do not intend to reply to the criticisms here, but I would ask my critics not to consider that due to inability or contempt. Perhaps an opportunity will present itself elsewhere.

THE AUTHOR

Contents

A SURVEY OF
PHYSICAL THEORY

The Unity of the Physical Universe

In endeavouring to claim your attention for a short time, I would remark that our science, Physics, cannot attain its object by direct means, but only gradually along numerous and devious paths, and that therefore a wide scope is provided for the individuality of the worker. One works at one branch, another at another, this one applies one method, that one another, so that the physical universe with which we are all concerned appears in different lights to different workers. Still, I hope to be able to claim your interest when I attempt to present to you the leading features of the physical universe as I see it in the observations and experiments at my disposal, and as it will probably be developed in the future.

I

As long as Natural Philosophy exists, its ultimate highest aim will always be the correlating of various physical observations into a unified system, and, where possible, into a single formula. For this, two different methods are available, methods which are often at variance with one another, but which more frequently mutually correct and assist one another, particularly when applied to the same purpose by one investigator. The earlier method considers isolated observations easily generalized, and places a single conception or theorem in a central position, and then tries to include in it, with more or less success, all the manifestations of Nature. It was in this way that Thales of Miletus, Wilhelm Ostwald, and Heinrich Hertz placed water, energy, and the principle of the most direct path respectively in the central position of the physical universe in which all physical facts are related and find an explanation.

The other method is more circumspect, less assuming and more reliable, but for a long time it had not the force of the first method, and so did not come into its own until much later. It renounced provisionally all claim to finàlity, and at first only followed those lines which appeared to have been completely established by direct experiment, leaving their further extension to later investigators. The best example of this is found in Gustav Kirchhoff's well-known statement of the function of mechanics, namely, that it is a description of all known motions. Each method supplements the other, and in no case can physical research ignore either.

However, I do not wish to talk to you about these duplicate methods of science. I would rather direct your attention to the more important questions as to whither these individual methods have led us, and to what goal they are likely to lead us. That physics in its development has really advanced, that we have learnt to know Nature appreciably better each decade cannot be seriously denied to-day by anybody. This is proved by a glance at the constant increase in the number and significance of the methods employed, methods which man investigates in order to make use of Nature for his own ends. But in what direction does this advance proceed as a whole? How far can we say that we are actually approaching the goal we seek, namely, the unified system? The answers to these questions must be of the greatest importance to every physicist who studies the progress of his science. When we are in a position finally to answer them we shall also be in a position to consider the broader question, much discussed to-day, as to what is the fundamental meaning of the so-called physical universe to us. Is it merely a practical, though fundamentally arbitrary, creation of our imagination, or are we forced to the opposite conception that it reflects real natural phenomena independent of us?

In order to realize in what direction the extension of physical science is moving, there is only one method of procedure. The conditions prevailing to-day must be compared with those of an earlier time. One asks further, what external criterion is possible in order to give the best characteristic of the state of development of a science? I can suggest nothing more general than the manner in which a science defines its fundamental conceptions and classifies its various branches. For the latest and most mature results of research must be embraced in rigid and convenient definitions, and by a suitable classification of subjects.

Let us now consider how physics has fared in this connection. We perceive at once that physical research in all its branches deals with practical requirements, or with particularly striking phenomena. The initial classification of physics and the designation of its individual branches are decided according to the point of view taken. Thus geometry arose out of geography or land surveying, mechanics from engineering, acoustics, optics, and heat from the respective sense-perceptions, electricity from the remarkable results obtained by rubbing amber; and the theory of magnetism from the striking properties of iron ore found near the town of Magnesia. Thus the physiological element affects all physical definitions if all our observations be associated with our sense-perceptions; in short, the whole range of physics, its definitions as well as its entire structure, bears, in a certain sense, an anthropomorphous character.

How different from this is the picture which modern theoretical physics presents to us! First of all, the whole gives us a much more uniform impression, the number of different divisions has been considerably diminished as the allied branches have been fused together. Thus acoustics is completely absorbed in mechanics, magnetism and optics in electro-dynamics, and these simplifications are accompanied by a striking severance from the human elements in all physical definitions. What modern physicist thinks of rubbed amber when considering electricity, or, when dealing with magnetism, thinks of the small Asiatic town where the first natural magnet was found? Further, the sense-perceptions have been definitely eliminated from physical acoustics, optics, and heat. The physical definitions of sound, colour, and temperature are to-day in no way associated with the immediate perceptions of the respective senses, but sound and colour are defined respectively by the frequency and wavelength of oscillations, and temperature is measured theoretically on an absolute temperature scale corresponding to the second law of thermo-dynamics, or, in the kinetic theory of gases, as the kinetic energy of molecular motion. In practice it is measured by the variation in volume of a thermometric substance, or by the deflection of a bolometer or thermocouple. It is in no way described as a feeling of warmth.

Exactly the same considerations apply to the conception of force. Originally, the word *force* without doubt referred to human force, corresponding to the use of men or beasts to work the first and oldest machines—the lever, the pulley, and the screw. This proves that the conception of force was derived at first from the sense of strength and muscle, that is, that it was a specific sense-perception. In the modern definition of force, however, the sense-perception is eliminated even as is the sense of colour from the definition of colour.

Originally, all branches of physics were more or less united through being associated with the senses. The elimination of the specific sense-element, however, was sufficient to break down the connecting bonds and divide these into several distinct branches, in direct contrast to the general tendency towards unity and assimilation. The theory of heat offers the best example of this process. Formerly heat constituted a certain well-defined single branch of physics, characterized by the sense of warmth. In all text-books of physics dealing with heat, we now find an entire sub-division—namely, heat radiation—is set apart and considered with optics. The significance of the sense of warmth is no longer sufficient to hold together the heterogeneous parts. Furthermore, one part is given to optics, i.e. electro-dynamics, and the other to mechanics, in particular, the kinematic theory of matter.

Looking back over the preceding, we may say briefly that the feature of the whole development of theoretical physics, up to the present, is the unification of its systems which has been obtained by a certain elimination of the anthropomorphous elements, particularly the specific sense-perceptions. Seeing, however, that the sensations are acknowledged to be the starting-point of all physical research, this deliberate departure from the fundamental premises must appear astonishing, if not paradoxical. Yet there is hardly a fact in the history of physics so obvious now as this, and, in truth, there must be undoubted advantages in such self-alienations.

Before we examine further this important point, we will direct our attention from the past and present to the future. Into what groups will physics be split up in years to come? At present there are two large divisions, namely, mechanics and electro-dynamics, or, as they may also be called, physics of matter and physics of the ether. The former includes acoustics, heat, and chemical phenomena; the latter embraces magnetism, optics, and heat radiation. Are these divisions final? I think not, since the line of demarcation between the two cannot be exactly determined. For example, do the phenomena of light emission belong to mechanics or to electro-dynamics; and in which category may the laws of motion of electrons be placed? Perhaps at first sight one would say they belong to electro-dynamics, since in the theory of electrons ponderable matter plays no part. But let attention be directed simply to the motion of free electrons in metals. There it will be found, from a study of the investigations of H. A. Lorentz, that the laws themselves fit in far better with the kinetic theory of gases than with electro-dynamics. Above all, it appears to me that the original difference between ether and matter is gradually disappearing. Electro-dynamics and mechanics are not nearly so distinct as is usually assumed in many places, and there is already some talk of a division into opposing camps of mechanical and electro-dynamical views of the universe. As fundamental in mechanics we need principally the conceptions of space, time, and motion, and it may be denoted by matter or condition. The same fundamentals are equally necessary to electro-dynamics. A slightly more generalized view of mechanics might thus allow it to include electro-dynamics, and, in fact, there are many indications that these two divisions, which are already encroaching upon one another, will be joined in one single general scheme of dynamics.

If the difference between ether and matter is overcome, on what definite grounds will the division of the system of physics be made? As we have seen above, the removal of this difference is the whole aim of the extension of science, and it is necessary that we consider more particularly the characteristics of physical principles before proceeding further.

II

To this end, I would direct your attention to that point from which the first step was taken towards the actual realization of the unified system of physics, up to then only postulated by philosophers—I mean to the Principle of Conservation of Energy. For the conception of energy, as well as the conceptions of space and time, is common to all the different regions of physics. Now, as I observed before, it will be clear, almost self-evident, to you that before its general enunciation by Mayer, Joule, and Helmholtz, the principle of energy originally had an anthropomorphous nature. It began with the knowledge that no one can do useful work out of nothing; and this knowledge was born of the observations gathered in the search for a solution of a technical problem—the discovery of perpetual motion. Thus perpetual motion became of as far-reaching importance in physics as the philosopher's stone in chemistry, though the advantages to science were derived from the negative rather than from the positive results of the experiments. To-day we talk of the principle of energy without any reference to the human or the technical point of view. We say that the total energy of an enclosed system of bodies is a quantity whose value cannot be increased or decreased by any reactions within the system. We no longer think of making the validity of this theorem dependent upon the refinements of methods which we possess at present to get experimental proof on the question of perpetual motion. In this generalization, which cannot be rigidly proved, but which forces itself upon our attention, lies the above-mentioned emancipation from the anthropomorphous elements.

While the principle of energy stands before us as a complete, self-contained picture, freed from and independent of the vicissitudes of its development, this is by no means the case, to the same extent, with that principle introduced into physics by R. Clausius as the Second Law of Thermo-dynamics. Indeed, the fact that this theorem has not yet been completely established gives it a particular interest in present-day discussions. The second law of thermo-dynamics still retains a particularly anthropomorphous character in current criticism. There are several distinguished physicists who connect the question of its validity with the inability of men to penetrate the molecule and make it similar to Maxwell's demon, which, without doing work, can separate the more rapid from the slower molecules of a gas merely by well-timed movements backwards or forwards on a small path. One need not be a prophet to be able to predict with certainty that the kernel of the second theorem has nothing to do with human ability, and that its final presentation must and will follow in a manner which bears no relation to the feasibility of performing any natural processes through human agency. I hope that the following discussion will contribute to this emancipation of the second law.

Let us next examine more closely the meaning of the second law, and its relation to the principle of energy. While the principle of energy limits natural phenomena in that it never permits energy to be created or destroyed, but only transformed, the second law increases the restraints in that it only permits certain kinds of transformations, and those only under certain conditions. Thus mechanical work can be transformed into heat without more ado; for example, by friction, but the converse is not true. Were the latter possible, then the heat of the earth, of which an unlimited supply is available, could be applied to drive a motor, and this would give the double advantage of using this motor as a refrigerator, since it would cool the surface of the earth.

On account of the practical impossibility of such a motor, which would be a case of perpetual motion of the second class, it necessarily follows that there are phenomena in Nature which cannot be made completely reversible. If, for example, friction, by means of which mechanical work can be transformed into heat, could be made completely reversible in some manner by means of some complicated apparatus, then this apparatus would be another example of the motor mentioned above—a case of perpetual motion of the second class. This is immediately evident if one considers what the apparatus would perform—

a transformation of heat into work without any intermediate variations. Let us call an operation, which cannot completely be made to go backwards in any way, an irreversible process, and all remaining operations reversible processes, we then get straight to the kernel of the second law of thermo-dynamics if we say that there are irreversible processes in Nature. Accordingly the changes in Nature have a definite direction. The world takes a step forward with every irreversible process, and this step can in no circumstances be completely retraced. In addition to friction, the following are examples of irreversible processes: Conduction of heat, diffusion, conduction of electricity, emission of light and heat rays, atomic disruption of radioactive substances, etc. Examples of reversible processes, on the other hand, are: Movements of planets, free fall in a vacuum, undamped oscillations of a pendulum, propagation of light and sound waves without absorption and refraction, undamped electrical oscillations, etc. All these operations are either periodic in themselves, or completely reversible by means of suitable apparatus, so that there is no change in the final state; for example, in the free fall of a body, by projecting it upwards with the velocity due to the fall, it will rise to its original height; and in the case of light or sound waves, by complete reflection at suitable surfaces.

Now what are the general characteristics of irreversible processes, and what is the general quantitative measure of irreversibility? These questions have been answered in many different ways, and the study of their history offers a specially interesting insight into the typical method of development of a general physical theory. Just as, originally, the principle of energy was arrived at as a result of the technical problem of perpetual motion, so also another technical problem, that of the steam engine, brought out the differences between reversible and irreversible processes. Sadi Carnot himself realized (although he made use of a theory of heat which has since been disproved) that the irreversible processes are less efficient than the reversible, or that in an irreversible process a certain amount will be wasted when converting heat to mechanical work. What more natural than to adopt as a measure of the irreversibility of a process the quantity of mechanical work lost? In a reversible process the amount of work definitely lost will then be nothing. This method of presentation has proved itself useful in certain special cases—for example, in isothermal processes, and is still of use at the present day. For general purposes, however, it has been shown to be unusable and even misleading. This is due to the fact that the amount of work lost in any particular reversible process cannot be definitely stated when the source of energy, which should have supplied this work, has not been particularized.

An example will make this clear. The conduction of heat is an irreversible process, or, as Clausius expresses it, heat cannot pass from a cold to a hot body without compensation. What is the work which is definitely lost when the (small) quantity of heat Q passes by direct conduction from a hot body, temperature T_1, to a cold body, temperature T_2? To answer this question we use the so-called heat transmission theory in order to introduce a reversible Carnot cycle between the two bodies as heat containers. It is known that thereby a certain quantity of work is obtained, and this amount of work is what we require, for it is lost even in the direct transmission of heat by conduction. However, this amount of work has no definite value until we know whence the work was derived, whether from the hot body, cold body, or from some other source. It might be thought that the heat given out by the hot body during the reversible cycle is not equal to that taken up by the cold body, since a certain amount of heat is transformed into work. With equal justification, the quantity of heat Q transmitted by direct conduction may be identified with the heat given out by the warm body or with that taken up by the cold body. According to whether the first or second alternative is adopted, the amount of work lost in the conduction is—

$$Q \frac{T_1 - T_2}{T_1} \text{ or } Q \frac{T_1 - T_2}{T_2}.$$

Clausius realized this uncertainty, and therefore generalized the simple Carnot cycle by assuming a third heat container whose temperature is quite undetermined, and accordingly yields an unknown amount of work.* We see, therefore, that the attempt to consider the irreversibility of a process mathematically does not in general succeed, and, at the same time, we see the particular reasons for its failure. The question has been too anthropomorphically set. It is too closely connected with human needs to give us an expression for useful work. If one wishes to obtain a definite answer from Nature one must attack the question from a more general and less selfish point of view. We will now endeavour to do that.

Let us consider any ordinary process in Nature. This brings all bodies concerned from an initial state A to a final state B. The process is either reversible or irreversible, a third alternative is not possible. Whether reversible or irreversible it depends simply and solely on the initial and final conditions and not on the intermediate states. The answer to the question will depend on whether, having once attained the state B, a complete return to A can or cannot be accomplished in any way. If a complete return from B to A is not possible, and the process therefore is irreversible, then the state B can evidently be expressed in terms of a certain characteristic of the state A. Years ago I took the liberty of describing it thus—that Nature has a greater predilection for the state B than for the state A. From this point of view, Nature does not permit those processes for whose final states she has a less predilection than she has for the initial states. Reversible processes are limiting cases. In them Nature has as much liking for the initial as for the final state, and the transition between them can be made arbitrarily in either direction.

Now it behoves us to seek a physical quantity whose value can serve as a general measure of Nature's predilection for any state. This quantity must be such that it must be defined immediately by the condition of the system under consideration, without necessitating any knowledge of the previous history of the system, as is the case with energy, volume, and other characteristics. These quantities would possess the peculiarity that they would be increased in all irreversible processes,

---

* R. Clausius, "Die mechanische Wärmetheorie," 2nd Ed., Vol. I, p. 96, 1876.

whereas they would remain unchanged in all reversible processes, and the change caused by a process would provide a general measure of the irreversibility of the process.

R. Clausius actually discovered this quantity, and called it *Entropy*. Every system of bodies has in every state a certain entropy, and this entropy denotes Nature's predilection for that particular state. Whatever takes place within the system the entropy can only increase. It cannot decrease. If processes be considered involving effects due to bodies outside the system, those bodies must be considered as belonging to the system, and then the theorem in the above form still holds. Thus the entropy of a system of bodies is simply equal to the sum of the entropies of the individual bodies, and the entropy of a single body is found, after Clausius, by means of a certain reversible cycle. Conduction of heat into a body increases its entropy by an amount equal to the quantity of heat introduced divided by the temperature of the body: simple compression, on the other hand, does not alter the entropy.

To return to the above-cited example of a warm body temperature T_1, cold body temperature T_2, and a direct conduction of heat Q, then, according to what we have already said, the entropy of the hot body is decreased in the process, whereas the entropy of the cold body is increased, and the sum of both changes, i.e. the change of total entropy of both bodies is—

$$- \frac{Q}{T_1} + \frac{Q}{T_2} > 0.$$

This positive quantity gives, independently of all arbitrariness, a measure of the irreversibility of the process of heat conduction. Such examples can naturally be multiplied indefinitely. Every chemical process contributes to them.

The Second Law of Thermo-dynamics, with all its extensions, has become the Principle of Increase of Entropy, and you will now understand why, in considering the question earlier, I laid such emphasis on the very important part that would be played by reversible and irreversible processes in the physics of the future.

In fact, all reversible processes, occurring in matter, ether, or both, show a much greater resemblance one to another than to any irreversible process. The consideration of the differential equations which govern them demonstrates this. In the differential equations of reversible processes the time differential always occurs in even powers only, corresponding to the condition that they are independent of the sign of the time. This holds equally for the oscillations of a pendulum, for electrical oscillations, sound and light waves, as for motions of particles and electrons, if all damping is excluded. To this class belong infinitely slow processes in thermo-dynamics which entail several equilibrium conditions in which time plays no part at all, or, otherwise expressed, appears to the power zero, and as such can be considered as an even power. All these reversible processes have the common property also, as Helmholtz has shown, that they can be completely represented by the Principle of Least Action, which gives the same answer independently of the path pursued, and, so far, the theory of reversible processes can be considered as completely determined. Reversible processes have the disadvantage that they, all and sundry, are ideal. In Nature there is not a single reversible process, since every natural operation more or less involves either friction or conduction of heat. The principle of least action is no longer applicable when dealing with irreversible processes, for the principle of increase of entropy introduces into physics an entirely new element foreign to the principle of least action, and which claims special mathematical treatment. The constant trend towards a fixed final state corresponds to this entropy principle.

I hope that the above considerations will have sufficed to make it clear that the difference between reversible and irreversible processes lies much deeper than that between mechanical and electrical processes. Therefore this contrast may, with more accuracy, be made the most convenient basis for partition of all physical phenomena, and may play the chief part in the physics of the future.

And yet the classification outlined needs to be improved materially. For it cannot be denied that, in the form discussed, physics is still permeated with strong doses of anthropomorphism.

In the definition of irreversibility, as in that of entropy, reference was made to the feasibility of certain changes in Nature, and that means fundamentally nothing more than making the division of physical phenomena dependent upon the ability to carry out experiments. These experiments will surely not always remain on the same level, but will always become more and more complete. If the difference between reversible and irreversible processes has a permanent meaning for all time, it must be deepened and made independent of human ability. How this can be done will be described below.

III

The original definition of irreversibility, as we have seen above, has the serious defect that it assumes a certain limit to human knowledge, whereas actually no such limit can be proved to exist. On the other hand, man makes every effort to extend the present limits of his powers, and we hope that, later on, much may be accomplished that to-day seems impossible. May it not happen that a process, considered up to the present to be irreversible, will prove to be reversible as a consequence of some new discovery or invention? The whole fabric of the second law of thermo-dynamics would then certainly collapse, for the irreversibility of one single process involves that of all the rest, as may easily be shown.

Let us take a concrete example. The very remarkable rapid movements, easily seen with a microscope, which small particles, suspended in a liquid, undergo—the so-called Brownian movement—is, according to the latest researches, a direct result of the continuous impact of molecules of the liquid on the particles. If, by means of very refined apparatus, but without expenditure of work, it were possible to influence the individual particles so that the disorderly movement becomes an orderly one, then, without doubt, it would be possible to transform part of the heat of the liquid into perceptible, and, therefore also, usable active force. Would not this contradict the second law of thermo-dynamics? If this could be done, the law would cease to rank as a principle, since its validity would depend on the advancement of experimental technique. Hence the only method of assuring the position of the second law is to make the conception of irreversibility independent of experiment.

The conception of irreversibility brings us back to the conception of entropy, for a process is irreversible if it is associated with an increase of entropy. Hence the problem resolves itself into finding a more suitable definition of entropy. According to Clausius's original definition, entropy is measured by a certain reversible process, and the weakness of this definition is due to the fact that many such reversible processes—indeed, all of them fundamentally—are impracticable. It might be argued with a certain amount of accuracy that this is hardly a question of actual experiments and an actual physicist, but of ideal processes, so-called ideal experiments and an ideal physicist, who applies all experimental methods with absolute accuracy. But this involves a further difficulty. How far can these ideal measurements by the ideal physicist be said to exist? That it is possible to compress a gas with a pressure equal to the pressure of the gas, and to heat it from a heat container whose temperature is equal to that of the gas, can be understood if suitable extensions of limits be assumed. But it must appear doubtful that, for example, saturated steam can be liquefied by isothermal compression along a reversible path without destroying the homogeneity of the substance, as is assumed in certain thermo-dynamical investigations. Much more striking is the trust displayed in ideal experiments by the theorist in physical chemistry. With his semi-permeable membranes which can only be realized in practice under certain quite special circumstances, and then only to a certain approximation, he separates along reversible paths not only all kinds of molecules, whether in a stable or unstable condition, but separates the oppositely charged ions from one another and from undissociated molecules. He ignores both the enormous electrostatic forces which oppose such a separation, and the circumstance that, in practice, at the commencement of the separation of the molecules the ions are partly dissociated and partly reunited. Such ideal processes are, however, very necessary to compare the entropy of the undissociated molecules with that of the dissociated molecules. Indeed, it is almost marvellous that all these bold ideas have been verified so well by experience. If one considers, however, that in all these results there is no question of the actual feasibility of these ideal processes—the relations considered involving only directly measurable quantities such as temperature, wavelength, etc.—then it cannot be proved that introducing ideal processes as above takes us far afield. Neither can it be proved that the real meaning of the principle of increase of entropy, with all its consequences, can be fully dissociated from the original conception of irreversibility and from the impossibility of perpetual motion of the second class, just in the same way as the principle of conservation of energy cannot be dissociated from the impossibility of perpetual motion of the first class.

Let us proceed another step. To complete the emancipation of the conception of entropy from human experimental methods, and thereby create a real principle of the second law, was the life-work of Ludwig Boltzmann. Briefly it consists in making the conception of entropy depend upon that of probability. Thus the meaning of the word I used above is explained, namely, the *predilection* of Nature for a certain state. Nature prefers more probable to less probable states, in that all natural processes tend to the formation of states of greater probability. Heat flows from a body of high temperature to a body of lower temperature, because the state of equal temperature is more probable than a state of unequal distribution of temperature.

The calculation of a certain measure of probability for every state of a system of bodies has been rendered possible by the introduction of the atomic theory, and of statistical methods. To determine the mutual action of individual atoms the known laws of dynamics, mechanics, and electro-dynamics can be used unaltered.

In view of these considerations the second law of thermodynamics is ousted immediately from its isolated position, the mystery of Nature's predilection vanishes, and the entropy principle, bound up with the introduction of the atom into physics, becomes a well-established theorem in probability.

Now it is not to be denied that this step towards unification has been accompanied by many sacrifices. The principal sacrifice is that we are denied the complete answer to all questions relating to details of operations in physics, but this is a direct consequence of employing statistical methods. For example, in the calculation of mean values we do not consider individual elements of which they are composed.

A second serious disadvantage appears to lie in the introduction of two different sorts of causal relations between physical states. On the one hand, absolute necessity; on the other, mere probability of their interdependence. If a heavy liquid at rest sinks to a lower level, a necessary consequence, according to the principle of the conservation of energy, is that it can only be put in motion—i.e. acquire kinetic energy—if its potential energy is decreased, that is, its centre of gravity lowered. However, it is only extremely probable, and not absolutely necessary, that a warm body give up its heat to a cold body in contact with it, for it is possible to devise special arrangements and velocity conditions of the atoms such that the converse could happen. From this Boltzmann has deduced that such peculiar phenomena which contradict the second law of thermo-dynamics could occur in Nature, and had therefore allowed room for it in his conception of the physical universe. In my opinion there is no need to follow him here. For a universe in which things can happen such as the return flow of heat into a hot body, or the separation of two interdiffused gases, would not be the same as our universe. So long as we are only dealing with the latter it will be more profitable to neglect such exceptional operations, and to look for the general condition, and make natural assumptions excluding from the beginning all phenomena which are contrary to experience. Boltzmann himself has formulated such a condition for the theory of gases. To state it quite generally, it is the *hypothesis of elementary disorder*, or, briefly expressed, the assumption that individual elements, with which statistics deals, are completely independent of one another. The introduction of this condition establishes again the necessity of all natural occurrences. According to the laws of probability, a direct consequence of the fulfilment of this condition is an increase of entropy, so that the essence of the second law of thermo-dynamics can be described as the principle of elementary disorder! Thus enunciated, the principle of entropy is as little likely to lead to contradiction as is the law of probability, which is founded on strictly mathematical bases.

In what way is the probability of a system related to its entropy? This can be deduced simply from the theorem that the probability of two systems independent of one another is equal to the product of the probabilities of the two systems ($W = W_1 W_2$). The entropy, however, is represented by the sum of the entropies ($S = S_1 + S_2$). Therefore the entropy is proportional to the logarithm of the probability ($S = K \log W$). This theorem leads on to a new method, far more extensive than those of ordinary thermo-dynamics: a method of calculating the entropy of a system in a given state. Thus the definition of entropy applies not only to conditions of equilibrium, as is almost universally the case in ordinary thermo-dynamics, but also to any dynamical state, and for the calculation of entropy it is no longer necessary to introduce (as Clausius had to do) a reversible process, the reality of which always seemed more or less doubtful, but this makes one independent of all human technique. Anthropomorphism is eliminated, and thus the second law of thermo-dynamics set upon a real base just as was the first.

The utility of the new definition of entropy has been illustrated not only in the kinetic theory of gases, but also in theory of the radiation of heat, since it has led to the establishment of laws which agree very well with experiment. That radiant heat possesses entropy follows from the fact that a body emitting heat rays loses heat, and therefore its entropy decreases. But since the total entropy of a system can only increase, part of the entropy of the whole system must be contained in the heat radiated. Hence every monochromatic ray has a certain temperature depending only on its brightness. It is the temperature of a black body which emits rays of the requisite intensity. The chief difference between the theory of radiation and the kinetic theory is that in radiating heat, the elements, the disorder of which defines their entropy, are not merely atoms as in gases, but the extremely numerous harmonic oscillations, of which every homogeneous radiation, such as light and heat rays, must be composed.

For the laws of heat radiation in free space, it must be pointed out that the constants involved, analogous to the gravitation constant, are of a universal character, since they are independent of any special substance, or any special body. Thus they can be used to determine units of length, time, mass, and temperature, and these units must necessarily retain their meaning for all time and for all extra-terrestrial and superhuman *kulturs*. It is known that this does not, by any means, hold for our ordinary weight units. Though these are usually described as absolute units, it must be borne in mind that they bear special relation to our present terrestrial life. The centimetre is derived from the present circumference of our planet, the second from its time of rotation, the gramme from water as the principal constituent of the earth's surface, and temperature from the characteristic points of water. The former constants, however, are such that the inhabitants of Mars, and indeed all intelligent beings in our universe, must encounter at some time—if they have not already done so.

I wish to point out here a most remarkable theorem on the nature of entropy, arising out of its being bound up with probability. It is known that the theorem used above, which states that the probability of two systems is equal to the product of the probabilities of each system, holds only when the two systems are independent of one another from the point of view of probability. When that is not so the probability is different. Therefore, we may presume that in certain cases the total entropy of two systems is different from the sum of the entropies of the two systems separately. That such cases actually occur in Nature has been proved by Max Laue. Two wholly, or partially, *coherent* light rays (which are emitted from the same source), are not, in the meaning of probability, independent of one another, since the oscillations of one ray are to some extent determined with those of the other. A simple optical arrangement can actually be set up, by means of which two coherent rays of any temperatures can be directly transformed into others having a greater temperature difference. Therefore Clausius's fundamental theorem, that heat cannot flow from a cold to a hot body without compensation, does not hold for coherent heat rays.

But the principle of the increase of entropy still holds good in this case. The entropy of the original rays is not, in this case, equal to the sum of the entropies of the separate rays, but is less.*

Similar considerations obviously apply to the question, already referred to, as to the ultimate transformation of the Brownian molecular movement into useful work. For any apparatus designed to direct and order the individual moving particles (whether technically practicable or not), will, as soon as it functions, be in a certain sense *coherent* with the movement of the particles. Therefore it would in no way contradict the second law if useful kinetic energy arose from its working. It is only necessary to bear in mind that the entropy of the molecular movements is not to be merely added to the entropy of the machine.

Such considerations show how carefully we must proceed when estimating the entropy of any system from the entropies of its constituents. It is strictly necessary, when dealing with any part of a system, first to ask whether it is possible that at any other place in the system there is a coherent part of the system. Otherwise phenomena apparently contradicting the entropy principle might occur in the case of the unexpected mutual action of two sub-systems. But if the two sub-systems have no mutual action the error due to neglecting their coherence would not be noticeable.

Is not this analogous to the unknown relationships existing in psychology? These relationships are peculiar consequences of coherence, but often remain concealed and can, therefore, be ignored without disadvantage. Under certain external combinations of circumstances, however, they can give rise to quite unexpected effects.

Were we to allow our fancy free play, we could demonstrate the possibility, unrealizable at present, that, perhaps at distances too great for us to comprehend, systems exist coherent with our universe, and that while those systems remain separate from it, our universe behaves normally, but as soon as mutual

* M. Laue, *Ann. d. Phys.*, **20**, 365, 1906; **23**, 1, 795, 1907; *Verh. d. Dtsch. Physik. Ges.*, **9**, 606, 1907; *Physik. Ztschr.*, **9**, 778, 1908.

effects occur, apparent exceptions (but only apparent) to the entropy principle can arise. Thus it might be possible to avert the threatened danger of complete dissipation of heat deduced from the second law of thermodynamics, without having to question its general validity. This danger has made many physicists and philosophers unsympathetic towards the second law. But even without this artificial expedient it appears to me, already, on account of the unlimited increase of our observations of the surrounding universe, that we need not worry about this danger. Many far more urgent problems await solution at present.

IV

I have endeavoured to explain briefly some of the fundamentals which the physics of the future will probably have to examine. If we review cursorily the changes our conception of the universe has undergone as science developed, and bear in mind the characteristics of this development as outlined above, we must admit that the picture of the future appears colourless and drab when compared with the glorious colouring of the original picture, tinted with the manifold needs of human life, and to which all the senses contributed their part. This comparison detracts greatly from the value of an exact science, and then comes the greater difficulty that it is quite impossible to ignore the evidence of the senses (since we cannot exclude the admitted source of all our experiences), and consequently we can have no direct knowledge of the absolute.

Why is it then that, in spite of its evident disadvantages, the picture of the future can hold its own against all the past? It is simply the *unity* of the picture: unity of all separate parts of the picture, unity of space and time, unity of all experimenters, nations, and *kulturs*.

If we look at the question closely the older system does not form a single picture, but rather a collection of miniatures, for every set of natural phenomena has its own special picture. These different pictures do not coalesce. One can be removed without prejudice to the others, which will be impossible in the future picture of the physical world. No single part of it can be omitted as superfluous, each is an indispensable part of the whole, and as such possesses a special meaning in relation to observed facts. Conversely, every observable physical phenomenon will, and must, find its own particular place in the picture. Herein lies the essential difference between the usual pictures which in certain, but not in all, ways have to agree with the original—a difference which I think has been too much overlooked, even among physicists. Occasional remarks are found even in recent literature on the subject, such as in applications of the theory of electrons, or in the kinetic theory of gases, that these theories claim to give only an approximate picture of actual facts. If this assertion were expressed thus— that one could not demand agreement of *all* consequences of the kinetic theory of gases with observed facts, it would lead to serious misunderstanding.

When R. Clausius, in the middle of last century, deduced from the kinetic theory of gases that the velocity of gas molecules at ordinary temperatures was of the order of hundreds of metres per second, it was agreed that two gases diffused very slowly into one another, and that local temperature fluctuations in gases only disappear very slowly. Then Clausius defended his hypothesis by saying that this was only an approximate representation of facts, and that too much was not to be expected from it, but he showed by calculation of the mean free path that the picture drawn by him actually agreed with physical observations in two specified cases. He well knew that if any single contradiction were established the new theory of gases would irrevocably lose its place in the physical world picture: and the same is true to-day.

The extent to which a principle is recognized depends on the establishment of these claims, which are important to the physical picture, independently of the good wishes of individual workers, of rationality, of time, even of the entire human race. The last observation will appear at first sight daring, if not absurd. But remembering, for example, our earlier conclusions relative to the physics of the Martians, it must be admitted, at least, that the generalization is one which is used daily when conclusions are drawn directly from observations and can never be proved by such observations. Therefore anyone who disallows their force and significance thereby cuts himself adrift from physical thought.

No physicist doubts the truth of the observation that a being, endowed with physical intelligence, and possessing means of detecting ultra-violet rays, would recognize these rays as being similar to the normally visible ones, though no one has seen either an ultra-violet ray or such a being. No chemist has any scruples about ascribing to solar sodium the same chemical characteristics as those of terrestrial sodium, although he can never hope to put a salt of solar sodium into a test tube.

These last considerations lead to an answer to the questions propounded above. Is the physical world simply a more or less arbitrary creation of the intellect, or are we forced to the opposite conclusion that it reflects phenomena which are real and quite independent of us? Expressed in a concrete form, can we rationally assert that the principle of conservation of energy was true, even when nobody could think about it, or that the heavenly bodies will move according to the law of gravitation, when our earth and all that is therein is in ruins?

If, looking back over the past, I give an affirmative answer to these questions, I am certain that this answer is, in a way, contradictory to a tendency of natural philosophy (recently introduced by Ernst Mach) which is in great favour in scientific circles. According to this, nothing is real except the perceptions, and all natural science is ultimately an economic adaptation of our ideas to our perceptions, to which we are driven by the fight for existence. The boundary between physical and psychical research is only practical and conventional. The only real elements of the world are the perceptions.*

If we connect the last theorem with what we have understood from our review of the actual development of physics, we must arrive at the proper conclusion, namely, that what characterizes this development is the continual elimination of the real elements of the world from the physical picture. Every conscientious physicist must take great care to distinguish between

* Ernst Mach, "Beiträge zur Analyse der Empfindungen," Jena, 1886, pp. 23, 142.

his own unique ideal picture and those of all others. If two colleagues, working independently on a particular research, obtain contradictory results, as occasionally happens, the physicist would be very wrong to conclude that at least one of the two must be in error. The contradiction might be ascribed to the differences of a two-sided picture, but I do not think that any physicist would ever accept such an explanation.

Meanwhile, I am quite prepared to admit that an enormous improbability does not differ much from an actual impossibility, but I would expressly emphasize that the arguments which have been directed from every side against the atomic hypothesis and the electron theory are unjustifiable and unwarrantable. I would even like to assert—and I know I am not alone in this—that atoms, little as we know of their actual properties, are as real as the heavenly bodies, or as earthly objects around us, and when I say that a hydrogen atom weighs $1 \cdot 6 \times 10^{-24}$ gr., the statement involves as much learning as the statement that the moon weighs 7×10^{52} gr. It is true that I cannot place a hydrogen atom in a scale-pan, nor can I even see it, but neither can I place the moon in a scale-pan, and as far as seeing it is concerned, it is known that there are invisible heavenly bodies whose mass can be measured, with more or less accuracy—the mass of Neptune was measured before ever an astronomer turned his telescope towards it. There is not in existence a method of physical measurement in which all knowledge dependent on induction is eliminated. The same is also true for direct weighing and a glance into a laboratory shows us the combined results of observation and abstraction which such an apparently simple measurement involves.

It still remains to ask how it comes about that Mach's theory has won so many disciples among students of Nature. If I am not mistaken it represents, fundamentally, a reaction against the proud expectations of previous generations associated with special mechanical phenomena following the discovery of the energy principle, such as one can find buried in the writings of Emil du Bois-Reymond. I will not say that these expectations have not brought about many far-reaching pieces of work of lasting value—I will mention only the kinetic theory of gases—

but taken all in all their importance has been exaggerated. Physics has renounced the development of the mechanics of the atom, on account of the introduction of statistics. Mach's positivism was a philosophic example of the inevitable disillusionment. He fully deserved his success for having found again, in the sense-perceptions, the only legitimate starting-point of all physical research, and this in the face of threatened scepticism, but he overshot his mark, for he lowered the standard of the physical world picture to that of the mechanical world picture.

I am firmly convinced that there is no contradiction in Mach's system, if correctly followed through. It seems to me that his meaning is, fundamentally, only formalistic. This is not the nature of science and he evades the most convenient criterion of all scientific research—the finding of a *fixed* world picture independent of the variation of time and people. Mach's principle of continuity does not offer any alternative, for continuity is not constancy.

The fixed unity of the world picture is, however, as I have endeavoured to show, the fixed goal which true science approaches through all its changes. In physics we may correctly state that the present picture has certain properties, and, although it varies its hue according to the individuality of the worker, no revolution in Nature or Man can obliterate these properties. This constancy, independent of all human and intellectual individuality, is what we call reality. Is there, for example, a serious physicist to-day who doubts the reality of the energy principle? On the contrary the acknowledgment of this reality is regarded as essential to a scientist.

Indeed, however confidently one lays the foundations of the world picture of the future, no general rules can be formulated. The greatest foresight is here desirable. There is a second consideration in connection with this question. What it simply and solely depends upon is the realization of a definite object even though unattainable. This object is—not the complete adaptation of our ideas to our perceptions, but—the complete liberation of the physical picture from the individuality of the separate intellects. This is, perhaps, a more accurate paraphrase of what I have called above, in order to avoid misunderstanding, "the liberation from the anthropomorphous elements," as if a picture could be independent of the painter: this is a contradiction in terms.

Still another argument, in conclusion, which perhaps has made more impression than all previous considerations on those who are disposed to set up the human economic point of view as being the only fruitful starting-point. As the great masters of exact research threw out ideas in science—as Copernicus removed the centre of the universe from the earth, as Keppler propounded the laws named after him, as Newton discovered general gravitation, as Huygens set forth the undulatory theory of light, as Faraday created the foundations of electro-dynamics—very many more can be cited—the economic point of view was the very last with which these men armed themselves in the war against inherited opinions and commanding authority. No—they were moved by their fixed belief in the reality of their picture, whether founded on an intellectual or a religious basis. In view of this unassailable fact, we cannot prove, for the present, that if Mach's principle of economy be once actually placed in the centre of our knowledge, the ideas of such prominent intellects would be disturbed, the flight of their fancy weakened, and thereby the advance of science fatally retarded. Would it not be more truly economical to give the principle of economy a somewhat less conspicuous position? You will already have seen, from the method of putting the question, that I am far from leaving out the consideration of economy in a higher sense, or from banishing it altogether.

We can even go a step further. These men did not speak of their world picture, but only of the world or Nature. Is there any recognizable difference between their "world" and our "world picture of the future"? Certainly not. For it is known—after Immanuel Kant—that there is no method of proving the existence of such a difference. The expression "world picture" has only become common for the sake of caution, in order that certain illusions are excluded from the start. Applying the necessary foresight, and knowing exactly that we mean by "world" nothing but the ideal picture of the future, we can, if we wish, substitute the single word and obtain a more realistic expression. This expression evidently recommends itself far more from the economical standpoint than Mach's positivism, which is fundamentally very complicated and difficult to grasp, and it is actually used now by physicists when they talk about science.

I spoke of illusions just now. It would certainly be a serious illusion on my part if I hoped that my remarks have carried general conviction, or even that they have been generally understood, and I shall very anxiously leave it with you. Surely much more will be thought and written concerning these questions, for theorists are numerous and paper is patient. Therefore we wish to emphasize unanimously and frankly what is acknowledged by us all, without exception, and must be admitted. That is, in the first place, scrupulousness in criticism of self, bound up with perseverance in the fight for true knowledge, and, in the second place, honourable regard, not to be shaken by misunderstandings, for the personality of scientific adversaries, and finally, a lasting confidence in the force of the Word which, for more than 1900 years, has given us an ultimate infallible test for distinguishing false prophets from true—"By their fruits ye shall know them!"

The Place of Modern Physics in the Mechanical View of Nature

As is well known, enormous changes have taken place in physics since the time of Neumann and Helmholtz. If Helmholtz were with us to-day, he would doubtless be amazed at many of the facts of physics of which he would hear. First of all there are the immense advances in experimental technique which have brought about these changes. The results obtained have been so unexpected in many respects, that one is inclined, nowadays, to consider as soluble problems which were unthought of a few decades ago, and hardly anything is considered impossible from a technical point of view. The advance brought about in practical work has been communicated to the theorists. They go to work with a keenness unknown in times gone by. No physical theorem is at present beyond doubt, all and every physical truth is considered disputable. It often seems almost as if theoretical physics is about to be plunged again into chaos.

The more complex new facts are, and the greater the multiplicity of new ideas introduced, the more urgent is the need for a method of consideration which will combine them. For, as certainly as the result of any particular experiment can only be confirmed by a suitable arrangement and interpretation of the research, so surely a working hypothesis of use in a wider sphere, and which helps to give a correct exposition, can only be formed by means of a suitable conception of the physical universe. This demand for a unified view of Nature is of importance, not only to physics, but to all natural sciences; since a change in the scope of physical principles must react on all the other natural sciences.

The conception of Nature which has been of greatest use to

physics is unquestionably the mechanical conception. If we bear in mind that this conception aims at explaining all qualitative differences ultimately in terms of motion, we may define the mechanical conception of Nature as the view that all physical phenomena can be completely reduced to movements of invariable and similar particles or elements of mass. At all events, that is what I am going to call the mechanical conception of Nature. Is this hypothesis fundamental and practical in the light of recent developments of physics?

From the earliest times there have been physicists and philosophers, who regarded this as self-evident; indeed, they considered this conception as a postulate of physical research. From their point of view, the problem for theoretical physics is simply to interpret all phenomena in Nature in terms of motion. On the other hand, certain thinkers with sceptical minds doubted the fundamental character of such a formulation of the question and considered the mechanical view of Nature to be too narrow to embrace the whole range of natural phenomena. It cannot be said that either of the two opposed schools has obtained a decisive advantage. In these days, for the first time there seems to be a prospect of a final decision, as the result of a far-reaching movement affecting the whole of theoretical physics—a movement of such a radical and revolutionary nature that its effect is felt far outside the range of pure physics in the allied sciences of chemistry and astronomy, and even penetrates into psychology. Following this movement, dissensions rage among scientists comparable with those which raged round the Copernican view of the universe. I will endeavour to show what has led to this revolution, and how the crisis consequent upon it can be settled.

The golden age of the mechanical conception was the last century. The first great impulse was given by the discovery of the principle of conservation of energy, indeed, the mechanical conception has sometimes been identified with the principle of energy, particularly at the time of its discovery. This misunderstanding was due to the fact that from the standpoint of the mechanical view the principle of energy is easily deducible; for if all energy is mechanical, the principle of energy is fundamentally none other than the law of *vis viva*, which had been known a long time. There are only two kinds of energy in Nature, kinetic and potential, and when considering any particular form of energy, e.g. heat, electricity, magnetism, it has simply to be decided whether it is kinetic or potential. This is the attitude adopted by Helmholtz in his first epoch-making paper on the conservation of momentum. A certain time elapsed before it was realized that nothing regarding the nature of energy follows from the law of conservation of energy. This was advocated from the beginning by Julius Robert Mayer, the discoverer of the mechanical equivalent of heat.

It was the development of the kinetic theory of gases that gave the special stimulus to the mechanical conception of Nature. This, most providentially, coincided with the direction along which chemical research had progressed. There, from a study of the differences between molecules and atoms, Avogadro's Law was obtained (as was also the most useful definition of a molecule of a gas), and this law was deduced directly from the kinetic theory of gases, provided that the kinetic energy of the molecule in motion was taken as the measure of the temperature. On the basis of the atomic ideas, the processes of dissociation and association, of isomerism and of optical activity of the molecule were explained in detail by mechanics as successfully as were the physical phenomena of friction, diffusion, and conduction of heat.

Of course, the final and most important question of how the differences between the chemical elements could be interpreted in terms of motion remained unsolved. But here, also, there was hope, for the periodic system of the elements seemed to indicate clearly that there was only one ultimate form of matter; and though Prout's hypothesis, that hydrogen is this ultimate form, has, in the meantime, become untenable, since atomic weights in general are not exact multiples of the atomic weight of hydrogen, yet there is always the possibility of choosing a still smaller primary atom as the common basis for all chemical elements, and so preserving the uniformity of the element.

For a while a serious danger appeared to threaten the atomic theory from the point of view of energy, arising from the growth of pure thermo-dynamics. If it had been known, as was shown above, that the mechanical conception is in no way necessary to the principle of energy, the second law of thermo-dynamics and its many applications to physical chemistry, would have given rise to a certain doubt concerning the atomic theory. General laws, such as those governing latent heats of evaporation and liquefaction, osmotic pressure, electrolytic dissociation, lowering of freezing-point and raising of boiling-point, could, with ease, be deduced completely and exactly from pure thermo-dynamics. From the atomic theory, however, these laws could only be derived with difficulty, and then only to a certain degree of approximation, particularly so when dealing with fluids and solids, to which cases the atomic theory had not been fully applied. The methods of thermo-dynamics, however, governed the three conditions of matter equally, indeed, some of its most remarkable results were obtained when dealing with fluids. Above all, the irreversibility of all natural processes was a serious trouble to the mechanical conception, since all processes in mechanics are reversible. It needed the comprehensive analysis and the inflexible optimism of a Ludwig Boltzmann not only to reconcile the atomic theory with the second law of thermo-dynamics, but also to explain for the first time the fundamental basis of the second law by means of the atomic theory. All these difficult questions were easily settled, or rather they did not exist for the adherents of pure thermo-dynamics, who did not recognize the necessity of reducing thermal and chemical energy to mechanical, but accepted the assumption that energy could take different forms—a circumstance which caused Boltzmann to remark that the kinetic theory of gases seemed to him to be going out of fashion. He would not have said this a few years later, for it was about that time that the kinetic theory of gases began to yield results at least as important as previous results.

At last pure thermo-dynamics attained its natural position. Since the second law only gave an inequality, it was only conditions of equilibrium which could be derived from it, in complete generality and exactness. Beyond this, in the consideration of physical or chemical phenomena, the second law could only give the sign of the variation with time, and could only yield qualitative results for such phenomena near the condition of equilibrium. It gave no definite quantitative values for velocities of reaction, and still less an insight into the details of these processes. Here no course is left open but to apply the methods of the atomic theory, and these methods have, throughout, been successful. They have been of particular importance in deducing the laws of ionization and for all phenomena in which electrons play a part. I will merely remark that dispersion, cathode and Röntgen rays and the whole domain of radio-activity can only be explained on the lines of a kinetic atomic theory.

Even in the original sphere of thermo-dynamics, viz. equilibrium, or stationary conditions, the kinetic theory has thrown light on certain problems which would otherwise have remained unsolved. It has made more comprehensible the emission and absorption of heat rays; indeed, in the elucidation of the so-called Brownian movement, it has provided the direct, so to speak concrete, justification for its existence, and this has been its supreme triumph. In brief, it can be said that, in heat, chemistry, and electron theory, the kinetic theory of the atom is no longer merely a working hypothesis, but a lasting and established theory.

How does the mechanical conception of Nature stand now? The atomic theory of matter and electricity is not enough, for it further requires a representation of *all* phenomena in Nature in terms of motions of simple particles.

The most elegant, and perhaps the final, attempt to express all natural phenomena in terms of motion, is contained in Heinrich Hertz's mechanics. In this, the search of the mechanical conception for a uniform world picture has been brought to a somewhat ideal completion. Hertz's mechanics does not really represent physics as it is, it is physics as it might be, a sort of confession of faith for physics. It draws up a programme of stately consistency and harmony, a programme surpassing all previous investigations directed to that end. Hertz is not satisfied with postulating that all Nature, from the mechanical point of view, can be completely explained by assuming movements of simple, similar particles, which build up the whole of the physical universe. He goes beyond Helmholtz's conservation of momentum, in so far as he eliminates from the beginning the difference between potential and kinetic energy, and thereby all problems involving the investigation of special types of energy. According to Hertz, there is not only one form of matter, the particle, but also only one form of energy, kinetic. All other types of energy, such as potential energy, electromagnetic, chemical, and thermal energy, are really kinetic energy of the motion of imperceptible particles, and these different forms are simply and solely due to natural relations between the positions and velocities of the particles considered. These relations do not in any way prejudice the validity of the principle of energy, since they affect only the directions of the movements and not the magnitudes of the kinetic energies just as the direction of a train is altered by curvature of the rails, but not its speed. All movements in Nature, according to Hertz, depend ultimately on the inertia of matter. A good example is furnished by the kinetic theory of gases, in which the elastic energy of stationary gas particles hitherto considered as potential is replaced by the kinetic energy of moving gas particles. This radical simplification of the postulates gives a wonderful simplicity and lucidity to the theorems of Hertzian mechanics.

Closer examination shows that the difficulties are not eliminated, but merely pushed aside, indeed removed into the almost inaccessible domain of experimental verification. Hertz himself may have felt this, for he has never made any attempt to give, in any particular simple case, the imperceptible motions introduced by him and the requisite relations, as Helmholtz did in the introduction to his paper. Even to-day, we have made no progress in this direction; on the contrary, we shall see that physics has developed along other lines, which not only lead us away from the Hertzian view, but away from the mechanical conception generally. Among the physical phenomena which have been most carefully investigated, there is a large group, which it appears to be impossible to reconcile with the mechanical view of Nature.

I will now consider the real stumbling-block of the mechanical theory, namely, light-ether. The attempts to represent light waves as motions of a substance consisting of very minute particles are as old as Huygens's undulatory theory, and exceedingly varied theories have been put forward from time to time to explain this mysterious medium. It must follow from the mechanical conception that where there is energy there must be motion, and where there is motion there must be something which moves. Therefore, as the existence of a material light-ether is a postulate in the mechanical conception, it must be vastly different from every other known substance, on account of its very low density, compared with the enormous elasticity it must have to allow for the huge velocity of light. According to Huygens, who considered light waves to be longitudinal, the light-ether could be considered as a fine gas, but after Fresnel, who was certain that the waves were transversal, ether must be looked upon as a solid body; since a gaseous ether could not possibly propagate transverse light waves. Attempts have often been made to explain transverse waves by friction, which exists in gases, but these explanations cannot be accepted, as neither absorption of light nor variation of velocity with colour can be proved to exist in the free ether. Thus it was necessary to assume a solid body having the special property that the heavenly bodies could move through it without experiencing any resistance. It was here that the difficulties began. Every attempt to apply the theory of the elasticity of a solid body to the light-ether required longitudinal waves which, in reality, do not exist, at least they could not be discovered, in spite of strenuous and varied methods applied to do so. It is only possible to get rid of these longitudinal waves by assuming infinitely large or infinitely small compressibility of the ether. But, with this assumption, it is impossible completely to satisfy the boundary conditions at the surface of separation of two different media.

I will not go into all the various more or less complicated assumptions that have been made to overcome these difficulties. I will merely refer to one serious symptom, usually accompanying unfruitful hypotheses and particularly in evidence in this problem. I refer to the introduction, into physics, of controversies which cannot be settled experimentally. A conspicuous example is the quarrel between Fresnel and Neumann about the relation between the vibration direction of a ray of polarized light and the plane of polarization. There is hardly any part of physics in which a question apparently insoluble has raised such a bitter controversy, all the available weapons of theory and experiment having been brought into play.

With the introduction of the electro-magnetic theory of light, it was realized that the question was of no importance—at all events, of no importance in a method wherein light is considered as an electro-dynamic phenomenon. Though the problem of explaining light waves mechanically remained unsolved, it was only postponed to await the solution of the much more general problem, that of expressing in terms of motion all electro-magnetic phenomena, both static and dynamic. In fact, the interest in this more general problem became ever greater with the development of electro-dynamics. One got nearer the solution by attacking the problem more comprehensively from general considerations, and the significance of light-ether was emphasized. For hitherto the light-ether was only of importance in the consideration of light waves, but it now became the medium by which all electro-magnetic processes were transmitted, at least in a pure vacuum.

All these attempts were fruitless—light-ether mocked all efforts at explaining it mechanically. So much became evident, that in one sense electric and magnetic energies were interrelated in the same way as kinetic and potential energies, and the question became, whether electric or magnetic energy should be considered as kinetic. In optics, the former would agree with Fresnel's theory, the latter with Neumann's. But the hope that an examination of the properties of static and stationary fields would furnish the necessary evidence to settle the question, insoluble in optics, was not realized. On the contrary, the difficulties were appreciably increased. All conceivable proposals and combinations were tried to obtain the foundations of the constitution of the light-ether. Of all physicists engaged on the problem, Lord Kelvin, right up to the time of his death, appears to have been the most active. It was shown to be impossible to explain electro-dynamic phenomena in free ether on a purely mechnical hypothesis—whereas the same phenomena were represented, by means of the Maxwell-Hertz differential equations, very simply and accurately in all details examined up to that time. The laws themselves were known in detail, but the mechanical explanation denied these simple laws completely and without question. At least, I do not think any serious objection will be raised in physical circles when I remark that the assumption of the complete validity of the simple Maxwell-Hertz differential equations excludes any possibility of explaining, on mechanical lines, electro-dynamic phenomena in the ether. This is not altered by the fact that Maxwell originally derived his equations with the help of the mechanical theories. It would not be the first time that a correct result has been obtained from ideas associated not quite correctly. Whoever wishes to-day to adhere to the mechanical exposition of electro-dynamical phenomena in free ether, must regard the Maxwell-Hertz equations as not quite exact, and correct them by introducing certain terms of a higher order. From first principles there can be no objection to this point of view, and an extensive field is opened for all kinds of speculations. On the other hand, it must be borne in mind that this view can only be established by experiment, and that such experiments must only be considered together with all previous work, fruitless though it may have been. I have already spoken of such experiments. There is one, however, which I have not mentioned, the most important of all, since it is independent of all detailed assumptions as to the nature of the light-ether.

Whatever one thinks of the constitution of the light-ether, whether it is considered continuous or discontinuous, the question always arises, if, when dealing with the motion of a transparent body, the ether enclosed must be considered as moving with the body, or remaining totally or partially at rest while the body moves. It may with certainty be answered that, at all events, the light-ether is not always completely carried along with the body, often practically not at all. For, in a gas in motion, such as a current of air, the velocity of light is markedly independent of the velocity of the gas, or if I may so express it, light travels just as fast against the wind as with the wind. This was shown by Fizeau in the middle of the last century by means of delicate interference experiments. We must, therefore, suppose that ether, through which light waves are moving, is not noticeably affected by the movements of the air, but remains at rest, in spite of them. If this is so, the next question is naturally: What is the magnitude of the velocity with which atmospheric air moves through ether?

This is a question which it has so far been impossible to answer in any single case by measurement. The air surrounding the earth moves, as a whole, with the earth, and thus has a velocity relative to the sun of about 30 km. per second, the direction varying continuously through the seasons. Though this velocity is only a ten-thousandth part of the velocity of light, from our present knowledge of optics methods can be devised which allow a velocity of this magnitude to be measured. Researches into the measurement of the motion of the earth relative to the light-ether occupy many pages in the annals of physics. But much ingenuity and experimental skill have failed to overcome the difficulties encountered. Nature remains silent and refuses to divulge the answer. It has not been possible to discern any trace of the effect of the motion of the earth on optical phenomena in the atmosphere. In this connection the most striking result is that of A. Michelson, who compared the velocity of propagation of light parallel to the direction of motion of the earth and at right angles to it. In this work, the relations are so extremely simple and the methods of measurement so extraordinarily delicate, that any effect of the motion of the earth must have been detected. Such an effect could not be discovered.

Considering this state of affairs, so very difficult and mysterious in theoretical physics, it is not unnatural to wonder whether it would not have been better to attack the problem from a quite different point of view. What has been the chief cause of the failure of all researches relating to the mechanical properties of the ether? Is there any physical meaning to the questions of the constitution, density, and elastic properties of the ether, longitudinal ether waves, the relation of the velocity of the ether to the plane of polarization, and the velocity of the atmosphere relative to the ether? If not, the efforts made to solve these problems are analogous to the attempts to attain perpetual motion. Thus we arrive at a critical point.

In his Königsberg lecture, Helmholtz laid special emphasis on the fact that the first step towards the discovery of the principle of energy had arisen from the question: What must be the relations between the forces of Nature if it is impossible to obtain perpetual motion? Similarly, it can rightly be asserted that the first step towards the development of the principle of relativity was coincident with the question: What relations must exist between the forces of Nature if it is impossible to ascribe to the ether any material properties? Do the light waves advance through space of their own accord, without the aid of a material medium? If so, the velocity of a body relative to the ether cannot be defined, let alone measured.

I need not say that with this view the mechanical conception is absolutely untenable. Whoever regards the mechanical conception as a postulate in physical theory, cannot be amicably disposed towards the theory of relativity. The unprejudiced will ask: Where does this principle lead us? It must be first understood that the above purely negative formulation of the new principle can only lead to results when combined with a positive basis founded on experiment. The most suitable basis consists of the Maxwell-Hertz equations of electro-dynamic phenomena in free ether, or, as we had better say, in a pure vacuum. The vacuum is the simplest of all conceivable media and in the whole realm of physics there are no relations known—excepting general principles—which deal with such delicate phenomena and appear to express them so accurately as these equations.

A new truth always has to contend with many difficulties; if it were not so, it would have been discovered much sooner. The chief difficulty in the theory of relativity lies in a far-reaching, even revolutionary result, necessitating the reconsideration of our conception of time. I should like to explain this important statement by means of a concrete example.

According to the principle of relativity, it is absolutely

impossible to find a general constant velocity of the solar system as a whole by means of any measurements within the system. Such a velocity, however great, could not be detected from any effects within the system. This law is already familiar in astronomy, it must also be true in physics. Every educated man knows that when he observes a phenomenon in any heavenly body, such as the sun, the event does not take place on the sun at the same moment as it is observed at the earth, but a certain time elapses between the event and its observation, namely, the time light takes to travel from the sun to the earth. If it is assumed that the sun and the earth are both at rest, this time is about eight minutes (we can here neglect the motion of the earth round the sun). If, however, the sun and the earth move with a common velocity in the direction from the earth to the sun, the earth moving towards the sun and the sun away from the earth with the same velocity, the time will be shorter. After leaving the sun, the light wave, which brings news of the occurrence from the sun to the earth, travels through space independently of the velocity of the sun, and meets the earth sooner than it would if the earth were at rest. Conversely, if the earth moved away from the sun, the sun following at a constant distance, the time between the occurrence and the observation of it would be longer.

The question therefore arises: How much time *actually* elapses between the occurrence of the event on the sun and its observation on the earth? This question is synonymous with the question as to what is the *actual* velocity of the sun and earth. As, according to the theory of relativity, no physical interpretation can be ascribed to the latter question, the same must apply equally to the former question. In other words, in physics a definition of time is only of importance when the velocity of the observer is taken into account.

It follows that time, as well as velocity, has a merely relative significance, and that when considering two mutually independent events at different places the terms *earlier* and *later* may even be reversed for two observers. This reasoning appears at first sight monstrous, even incomprehensible, but perhaps not more so than would have seemed 500 years ago the assertion that what we call the vertical is not an absolute constant, but describes a cone in space every twenty-four hours. However right it may be in many cases to follow intuition, there are circumstances in which this course would be a serious check to the introduction of new great ideas into science. It is true that many fruitful ideas in physics arose from the obvious, but there are always some, not the least valuable, which had to override accepted ideas before being established.

We can all remember how, when we were children, we had great difficulty in realizing for the first time that there were people on the earth whose feet pointed towards ours, and that they walked on the earth as safely as we did, without falling off and without any discomfort. If anyone to-day were to put forward the argument that the lack of clearness is a material objection to the relative character of all space directions, he would simply be laughed at. I am not sure that the same will not be the case 500 years hence with anybody who doubts the relative nature of time.

The measure of the value of a new hypothesis in physics is not its obviousness but its utility. If the hypothesis proves successful, one becomes used to it and it gradually becomes obvious of its own accord. When the exploration of electro-magnetic effects was yet incomplete, it was thought impossible to visualize galvanic streams, electro-motive force and magnetic lines of force without the aid of flow in liquids, hydraulic pumps, and stretched strings. Now, the majority of students of electric-technics scorn the use of these incomplete analogies and prefer to work directly with electro-magnetic conceptions which have become reliable through use. I have occasionally noticed that conversely, attempts have been made to illustrate complicated fluid motions, such as Helmholtz's vortex motion, by means of electro-magnetic analogies.

What is the position with regard to the theory of relativity? It certainly makes very far-reaching claims on abstract physics, but its methods are convenient and universal, and, above all, give definite results which are comparatively simply formulated. Of the pioneers in this work we must mention first Hendrik Antoon Lorentz, who discovered the principle of relativity of time, and introduced it into electro-dynamics without making any important deductions from it. Then comes Albert Einstein, who boldly proclaimed it as a universal postulate, and Hermann Minkowski, who succeeded in setting forth the theory of relativity in mathematical terms.

It is evidently not an accident that these abstract problems have chiefly attracted the attention of mathematicians, especially after it was seen that the mathematical methods employed were the same as those for four dimensional geometry. But genuine unprejudiced experimental physicists are in no way opposed to the theory of relativity. For the present they allow matters to proceed quietly and take their stand from results that can be proved experimentally. It must be pointed out in this connection that there are many physical deductions from the theory of relativity, but that their verification demands measurements so accurate that the instruments used are taxed to the utmost limit of their sensitivity. This is due, in the first place, to the fact that the velocities of the bodies considered are usually extremely small compared with the velocity of light. The most rapid movements are encountered when dealing with electrons, and, therefore, the first positive results are to be expected in the dynamics of electrons. Meanwhile, the sensitivity of the instruments will be improved as time goes on, the accuracy of the measurements increased and the verification of the theory more exact. The position is exactly the same as it was when the figure of the earth was first in question. If the radius of the earth had not been so great compared with the lengths used in investigations, the shape of the earth and the relativity of all directions in space would have been recognized much earlier.

The significance of the analogy I have drawn between time and space goes much further. It is more than an analogy, it is an identity, at least in the mathematical sense. Minkowski's greatest work was to show that if time is measured in suitable imaginary units, the three dimensions of space and this time dimension appear absolutely symmetrical in the fundamental equations of physics. The transition from one direction in space to another is accordingly mathematically and physically equivalent to a transition from one velocity to another, and the theory of the relativity of all velocities is but an extension of the theory of relativity of all space directions. As the latter theory had a hard struggle before being generally recognized, so also the former will have to overcome much opposition, but no longer, as earlier, associated with danger to life and limb. The best means, indeed the only means, of coming to a decision lies in a closer investigation of the results to which the new ideas lead, and the following remarks should be regarded in that light.

According to the principle of relativity, the physical universe, as we know it, has four equally important interchangeable dimensions. Three of them we call space co-ordinates, the fourth we call time, and from every physical law can be derived three others by interchanging the co-ordinates.

The chief law of physics, the pinnacle of the whole system is, in my opinion, the *principle of least action*, which involves the four universal co-ordinates in a perfectly symmetrical form.* From this central principle, four equally important principles radiate symmetrically in four directions, corresponding to the four universal dimensions. Corresponding to the three space dimensions is the (threefold) principle of momentum, and to the time dimension the principle of energy. It was never before possible to follow these principles back to their common origin.

This conception sheds a new light on the relation of the mechanical to the energy conception of Nature, for the energy view rests on the principle of energy, while the mechanical view rests on the principle of momentum. The three well-known Newtonian equations of motion are nothing but the expression of the principle of momentum as applied to a material point. According to them, the change of momentum is equal to the impulse of the force, whereas, according to the principle of energy, the change of energy is equal to the work done by the force. Each of the two conceptions of Nature, the mechanical

---

* Since the principle of least action is usually expressed by means of a time integral, the time co-ordinate seems to be more in evidence than the other co-ordinates. This one-sidedness is, however, only apparent and is due to the method of expressing the integral, for the "action integral" (the function whose variation vanishes) of any physical phenomenon is an invariant for any Lorentz transformation.

and the energy, suffers from a certain one-sidedness. The first is essentially superior to the second in so far as it gives three equations, corresponding to the vectorial character of momentum, while the principle of energy furnishes only one equation. Of course, these remarks apply, not only to the motion of a particle, but to every reversible phenomenon in mechanics, electro-dynamics, and thermo-dynamics.

The effective mass of a moving body can be deduced from its momentum or its energy, and naturally, when considered from this point of view, loses its fundamental character and becomes of secondary importance. As a matter of fact, this method shows that the effective mass of a body is not constant, but is dependent on its velocity, and changes so that, when the velocity of the body approaches the velocity of light, the effective mass increases beyond all bounds. Therefore, according to the theory of relativity, it is absolutely impossible for a body to have a velocity as great as, or greater than, the velocity of light. Moreover, the effective mass of a body is not constant, but, strictly speaking, depends on its temperature. This follows, quite apart from the theory of relativity, from the fact that every body contains a certain quantity of radiant heat, varying with its temperature. The magnitude of this was first recognized by Fritz Hasenöhrl.

The question must now be asked, if the particle loses the properties of constancy, and invariability, hitherto generally regarded as fundamental, what is the real substance, what are the invariable elements from which the whole universe is built up? One may answer that the invariable elements of the system of physics based on the theory of relativity are the so-called *universal constants*: chiefly the velocity of light in a vacuum, the electric charge and mass of an electron at rest, the "elementary quantum of action" obtained from heat radiation and which probably plays a fundamental part in chemistry, the constant of gravitation, and many others. These quantities have a real meaning, since their values are independent of the condition, standpoint, and velocity of the observer. For the rest it must be remembered that there are still many details to be explained. Were we in a position to answer all these questions satisfactorily, physics would cease to be an inductive science, and it will certainly always be that.

It will be seen from these remarks that the principle of relativity does not merely disorganize and destroy, but in a much higher degree organizes and constructs—it simply discards a form which, owing to the advance of science, was already out of date. In place of the old confined structure, it erects a new one, more comprehensive and more lasting, and includes all the treasures of the former, naturally everything I have mentioned above, in a different, clearer grouping, and yet has room for discoveries still to be made. It shuts out of the physical universe those unessential factors introduced by the contingencies of human views and customs, and purges physics of its anthropomorphic elements, which arose from individual peculiarities of physicists, and the complete exclusion of which, as I have endeavoured to show elsewhere, is the real end of all physical knowledge. It opens to the progressive mind a perspective of almost immeasurable breadth and height, and leads him to co-ordinate results in a way unthought of in former periods, and beyond even the perfect mechanics of Heinrich Hertz. No one who has studied the lines of these new methods can long resist the spell cast by them, and it is quite conceivable that in a well-ordered mind, like that of Hermann Minkowski, it can arouse intense enthusiasm.

Physical questions, however, cannot be settled from æsthetic considerations, but only by experiment, and this always involves prosaic, difficult, and patient work. The important physical significance of the theory of relativity lies in the fact that it provides a precise answer, determined by experiment, to a series of physical questions, which earlier were complete mysteries. Therefore the principle must be regarded at least as a working hypothesis of great potentiality, particularly as opposed to the mechanical hypothesis of the light-ether. At present, the dispute centres about the dynamics of the electron. This question has been brought within the range of exact observation by the discovery of the deflection of free electrons by electric and magnetic influences. Experienced minds and skilful hands are now at work in various laboratories, independently of one another, and the result is awaited all the more anxiously because it appeared at first that observations contradicted the theory of relativity, while at present the balance seems to be leaning in favour of the principle.

The eyes of numerous physicists and friends of physics are on this fundamental research, and this society* has shown its interest in it by devoting a part of the revenue of the Trenkle Foundation towards such experimental work. We hope that it will produce a valuable contribution to the solution of this problem.

Now, whatever the ultimate decision will be—whether the principle of relativity is accepted or whether it is abandoned, whether we stand actually upon the threshold of an entirely new view of Nature or whether this advance will fail to lead us out of the darkness—clearness must, under all circumstances, be obtained, and to obtain it no price is too high. Even a disillusionment, if only it be fundamental and conclusive, is a step forward, and the sacrifices involved in the abandonment of the theory would be amply compensated for by the discovery of the treasures of new knowledge. I wish these words to be interpreted in the spirit of this society, for it can always be counted to its credit that it has never favoured any prejudiced line of advance in science, but has always decisively rejected work which tended in that direction. We can have no doubt that this will be so in the future and that we will always advance unceasingly, undeterred by the nature of results, into the light of truth.

* The Assembly of German Science and Medicine.

New Paths of Physical Knowledge

Experimental research in physics has not experienced, for a long time, such a stormy period, nor has its significance for human culture ever been so generally acknowledged as nowadays. Wireless waves, electrons, Röntgen rays, radio-active phenomena more or less arouse everybody's interest. If we consider only the wider question of how these new and brilliant discoveries have influenced and advanced our understanding of Nature and her laws, their importance does not appear, at first sight, to be commensurate with their brilliance.

Whoever tries to judge the state of present-day physical theories from a detached point of view may easily be led to the opinion that theoretical research is complicated to a certain extent by many new experimental discoveries, some of which were quite unforeseen. He will find that the present time is an unedifying period of aimless groping, in direct contrast to the clearness and certainty characteristic of the recent theoretical epoch, which may, therefore, with some justification, be called the classical epoch. Everywhere, old ideas, firmly rooted, are being displaced, generally accepted theorems are being cast aside and new hypotheses taking their place. Some of these hypotheses are so startling that they put a great strain on our comprehension and on our scientific ideas, and do not appear to inspire confidence in the steady advance of science towards a fixed goal. Modern theoretical physics gives one the impression of an old and honoured building which is falling into decay, with parts tottering one after the other, and its foundations threatening to give way.

No conception could be more erroneous than this. Great fundamental changes are, indeed, taking place in the structure

45

of theoretical physics, but closer examination shows that this is not a case of destruction, but one of perfection and extension, that certain blocks of the building are only removed from their place in order to find a firmer and more suitable position elsewhere, and that the real fundamentals of the theory are to-day as fixed and immutable as ever they have been. After this more general consideration we will examine thoroughly the basis of these remarks.

The first impulse towards a revision and reconstruction of a physical theory is nearly always given by the discovery of one or more facts which cannot be fitted into the existing theory. Facts always form a central point about which the most important theories hinge. Nothing is more interesting to the true theorist than a fact which directly contradicts a theory generally accepted up to that time, for this is his particular work.

What is to be done in such a case? Only one thing is certain: The existing theory must be altered in such a way that it is made to agree with the newly-discovered fact. But it is often a very difficult and complicated question to decide in what part of the theory the improvement has to be made. A theory is not formed from a single isolated fact, but from a whole series of individual propositions combined together. It resembles a complicated organism, whose separate parts are so intimately connected that any interference in one part must, to some extent, affect other parts, often apparently quite remote, and it is not always easy to realize this fully. Thus, since each consequence of the theory is the result of the co-ordination of several propositions, any erronous result deduced from the theory can generally be attributed to several of the propositions and there are almost always numerous ways out of the difficulty. Usually, in the end, the question resolves itself into a conflict between two or three theorems, which were hitherto related in the theory, at least one of which theorems must be discarded on account of the new facts. The dispute often rages for years, and the final settlement means not only the exclusion of one of the theorems considered, but also, and this is specially important, quite naturally the corresponding strengthening and establishment of the remaining accepted propositions.

The most important and remarkable result of all such dis-
putes arising in recent times is that the great general principles
of physics have been established. These are the principle of
conservation of energy, the principle of conservation of momentum, the principle of least action, and the laws of thermo-
dynamics. These principles have, without exception, held the
field and their force has increased appreciably in consequence.
On the other hand, the theorems that failed to survive are ones
which certainly served as apparent starting-points for all
theoretical developments, but only because they were thought
to be so self-evident that it was usually considered unnecessary
to mention them specifically, or they were completely forgotten.
Briefly, it can be said that the latest developments of theoretical
physics were vindicated through the triumph of the great
principles of physics over certain deeply rooted assumptions and
conceptions, which were accepted from habit.

A few such propositions may be mentioned in order to illustrate this. These had been accepted as self-evident foundations
of their respective theories without any consideration, but in the
light of new discoveries have been proved untenable or, at
least, highly improbable, as opposed to the general principles of
physics. I mention three: the invariability of the chemical atom,
the mutual independence of space and time, the continuity of
all dynamical effects.

It is naturally not my intention to recount all the weighty
arguments which have been directed against the *invariableness of
the chemical atom*; I will only quote a single fact which has led to
an inevitable conflict between a physical principle and this
assumption, which had always hitherto been considered as self-
evident. The fact is the continual development of heat from a
radium compound, the physical principle is conservation of
energy, and the conflict finally ended with the complete victory
of this principle, though at first many wished to throw doubt
on it.

A radium salt, enclosed in a lead chamber of sufficient thick-
ness, continually develops heat at a calculated rate of about
135 calories per grain of radium per hour. Consequently, it
always remains, like a furnace, at a higher temperature than its

surroundings. The principle of conservation of energy states that the observed heat cannot be created out of nothing, but that there must be a corresponding change as the cause. In the case of a furnace, we have the continual burning, in the case of the radium compound, through lack of any other chemical phenomenon, a variation of the radium atom itself must be assumed. This hypothesis has established itself everywhere, although it appeared daring when looked at from the previously accepted point of view of chemical science.

Strictly considered, there is, indeed, a certain contradiction in the conception of a variable atom, since atoms were originally defined as the invariable ultimate particles of all matter. Accordingly, to be quite accurate, the term "atom" should be reserved for the really invariable elements, as, perhaps, electrons and hydrogen. But apart from the fact that perhaps it can never be established that an invariable element exists in an absolute sense, such an uncertainty of meaning of terms would lead to a terrible confusion in the literature. The present chemical atoms are no longer the atoms of Democritus, but they are accurately determinable numerically by means of another and much more rigid definition. Only these numbers are referred to when atomic change is mentioned, and a misunderstanding in the direction indicated appears quite impossible.

Until recently the *absolute independence of space and time* was considered no less self-evident than the invariableness of the atom. There was a definite physical meaning to the question of the simultaneity of two observations at different places, without any necessity to inquire who had taken the time measurement. To-day it is quite the opposite. For a fact which up to now has been repeatedly verified by the most delicate optical and electro-dynamical experiments has brought that simple conception into conflict with the so-called principle of constant velocity of light, which came into favour through Maxwell and Lorentz, and which states that the velocity of propagation of light in empty space is independent of the motion of the waves. This fact is briefly, if not also clearly, called the relativity of all motion. If one accepts the relativity as being experimentally proved, one must abandon either the principle of constant velocity of light or the mutual independence of space and time.

Let us take a simple example. A time signal is sent out from a central station such as the Eiffel Tower by means of wireless telegraphy, as proposed in the projected international time service. Then all the surrounding stations which are at the same distance from the central station, receive the signal at the same time and can adjust their clocks accordingly. But such a method of adjusting time is inadmissible if one, mindful of the relativity of all motion, shifts his viewpoint from the earth to the sun, and thus regards the earth in motion. For, according to the principle of constant velocity of light, it is clear that those stations which, seen from the central station, lie in the direction of the earth's motion, will receive the signal later than those lying in the opposite direction, for the former move away from the oncoming light waves and must be overtaken by them, while the latter move to meet the waves. This, according to the principle of constant velocity of light, makes it quite impossible to determine time absolutely, i.e. independently of the movement of the observers. The two principles cannot exist side by side. So far, in the conflict, the principle of constant velocity of light has decidedly had the upper hand, and it is very probable that no change will come about, in spite of many ideas recently promulgated.

The third of the above-mentioned propositions deals with the *continuity of all dynamical effects*, formerly an undisputed hypothesis of all physical theories, which was condensed in Aristotle's well-known dogma: *natura non facit saltus*. But even in this stronghold, always respected from ancient times, modern research has made an appreciable breach. In this case, recent discoveries have shown that the proposition is not in agreement with the principles of thermo-dynamics, and, unless appearances are deceptive, the days of its validity are numbered. Nature certainly seems to move in jerks, indeed of a very definite kind. To illustrate this more clearly, may I present a straightforward comparison.

Let us consider a sheet of water in which strong winds have produced high waves. After the wind has completely died down, the waves will continue for an appreciable time and move from one shore to the other. But a certain characteristic change will take place. The energy of the big, long waves will be transformed, to an ever-increasing extent, into the energy of small short waves, particularly when beating against the shore or some other rigid obstacle. This process will continue until finally the waves have become so small, and the movements so fine that they become quite invisible to the eye. This is the well-known transformation of visible motion into heat, molar into molecular movements, ordered movements into disorder; for in the case of ordered movement, many neighbouring molecules have the same velocity, while in disorderly movements, any particular molecule has its own special velocity and direction.

The process described here does not go on indefinitely, but is limited, naturally, by the size of the atom. The movement of a single atom, considered by itself, is always orderly, since all the individual parts of an atom move with a common velocity. The bigger the atom, the less the dissipation of energy that can take place. So far everything is clear, and the classical theory fits in perfectly.

Now let us consider another, quite analogous phenomenon, dealing, not with water waves, but with rays of light and heat. We here assume that rays emitted by a bright body are condensed into a closed space by suitable reflection and are there scattered by reflection from the walls. Here, again, there is a gradual transformation of radiant energy of long waves into shorter ones, of ordered waves into disorderly ones; the big, long waves correspond to the infra-red rays, the small, short ones to the ultra-violet rays of the spectrum. According to the classical theory, one would expect all the energy to be concentrated ultimately at the ultra-violet end of the spectrum. In other words, that the infra-red and visible rays are gradually lost and transformed into invisible and chemically active ultra-violet rays.

No trace of such a phenomenon can be discovered in Nature. The transformation sooner or later reaches quite a definite, determinative end, and then the radiation remains completely stable.

Different attempts have been made to bring this fact into line with classical theory, but hitherto these attempts have always shown that the contradiction struck too deeply at the roots of the theory to leave it undisturbed. Nothing remains but to reconsider once more the foundations of the theory. And it has been once more substantiated that the principles of thermo-dynamics are unshakable. For the only way found hitherto which appears to promise a complete solution of the riddle, is to start with the two laws of thermo-dynamics. It joins these, however, with a new peculiar hypothesis, the significance of which in the two cases mentioned above, can be described as follows.

In the water waves, the dissipation of kinetic energy comes to an end on account of the fact that the atom retains its energy in a certain way, such that each atom represents a certain finite quantum of matter which can only move as a whole. Similarly, although the light and heat radiation is of a non-material nature, certain phenomena must occur, which imply that radiation energy is retained in certain finite quanta, and the shorter the wave length, and the quicker the oscillations, the more energy is retained.

Nothing can as yet be said with certainty of the dynamical representation of such quanta. Perhaps one could imagine quanta occurring in this manner, viz. that any source of radiation can only emit energy after the energy has reached a certain value, as, for example, a rubber tube, into which air is gradually pumped, suddenly bursts and discharges its contents when a certain definite quantity of air has been pumped in.

In all cases, the quantum hypothesis has given rise to this idea, that in Nature, changes occur which are not continuous, but of an explosive nature. I need only mention that this idea has been brought into prominence by the discovery of, and closer research into, radio-active phenomena. The difficulties connected with exact investigations are lessened, since the results obtained on the quantum hypotheses agree better with observation than do those deduced from all previous theories.

But, further, if it is advantageous for a new hypothesis that it proves itself useful in spheres for which it was not intended, the quantum hypothesis has a great deal in its favour. I will here just refer to one point which is particularly striking. When air, hydrogen and helium were liquefied, a rich field of experimental research in low temperatures was opened, and has already yielded a series of new and very remarkable results.

To raise the temperature of a piece of copper by one degree, from $-250°$ to $-249°$, the quantity of heat required is not the same as that necessary to raise it from $0°$ to $1°$, but is about thirty times less. The lower the initial temperature of the copper, the less is the heat necessary, without any assignable limits. This fact not only contradicts our accepted ideas, but also is diametrically opposed to the demands of the classical ideas. For though man had learnt, more than a hundred years ago, to differentiate between temperature and quantity of heat, the kinetic theory of matter led to the deduction that the two quantities, though not exactly proportional, moved more or less parallel to one another.

The quantum hypothesis has fully explained this difficulty and, at the same time, has given us another result of great importance, namely, that the forces produced in a body by heat vibrations are of exactly the same type as those set up in elastic vibrations. With the help of the quantum hypothesis, the heat energy at different temperatures of a body containing atoms of one sort only can be calculated quantitatively from a knowledge of its elastic properties—the classical theory was very far from accomplishing this. A large number of further questions arise from this, questions which at first sight seem very strange, such as, for example, whether the oscillations of a tuning-fork are absolutely continuous, or whether they follow the nature of quanta. Indeed, in sound waves, the energy quanta are extremely small, on account of the relatively small frequency of the waves; in the case of a', for example, they are of the order of three quadrillionths of work units in absolute mechanical units. The ordinary theory of elasticity would, therefore, need just as little alteration on account of this, as it needs on account of the circumstance that it looks on matter as completely continuous, whereas, strictly speaking, matter is of an atomic, that is, of a quantum nature. But the revolutionary nature of the new development must be evident to everybody, and though the form of the dynamical quanta may for the present remain rather mysterious, the facts available to-day leave no doubt as to their existence in some form or other. For that which can be measured, exists.

Thus, in the light of modern research, the physical world-picture begins to show an ever-increasing connection between its separate parts and at the same time a certain definite form: the refinement of the parts appeared to have been missing from the earlier less detailed view and must have remained concealed. But, the question can always be repeated afresh: How far does this progress fundamentally satisfy our desire for knowledge? By refining our world-picture, do we attain a clearer understanding of Nature itself? Let us now consider briefly these important questions. Not that I shall say anything essentially new about this matter, with its manifold aspects, but because at present opinions are so divided on some things, and because all who take a deep interest in the real aims of science must take one side or the other.

Thirty-five years ago Hermann von Helmholtz reached the conclusion that our perceptions provide, not a representation of the external world, but at most only an indication thereof. For we have no grounds on which to make any sort of comparison between the actualities of external effects and those of the perceptions provoked by them. All ideas we form of the outer world are ultimately only reflections of our own perceptions. Can we logically set up against our self-consciousness a "Nature" independent of it? Are not all so-called natural laws really nothing more or less than expedient rules with which we associate the run of our perceptions as exactly and conveniently as possible? If that were so, it would follow that not only ordinary common sense, but also exact natural research, have been fundamentally at fault from the beginning; for it is impossible to deny that the whole of the present-day development of physical knowledge works towards as far-reaching a separation as possible of the phenomena in external Nature from those in human consciousness.

The way out of this awkward difficulty is very soon evident if one continues the argument a step further. Let us assume that a physical world-picture has been discovered which satisfies all claims that can be made upon it, and which, therefore, can represent completely all natural laws discovered empirically. Then it can in no way be proved that such a picture in any way represents "actual" Nature. However, there is a converse to this, and far too little emphasis is usually laid on it. Exactly similarly, the still more daring assertion cannot be disproved, that the assumed picture represents quite accurately actual Nature in all points without exception. For, to disprove this, one must be able to speak of actual Nature with certainty, which is acknowledged to be impossible.

Here yawns an enormous vacuum, into which no science can penetrate; and the filling up of this vacuum is the work, not of pure, but of practical reason; it is the work of a healthy view of the world.

However difficult it may be to prove scientifically such a view of the world, one can build so well on it, that it will stand unperturbed by any assaults, so long as it is consistent with itself and in agreement with the observed facts. But one must not imagine that advance is possible, even in the most exact of all natural sciences, without some view of the world, i.e. quite without some hypotheses not capable of proof. The theorem holds also in physics, that one cannot be happy without belief, at least belief in some sort of reality outside us. This undoubting belief points the way to the progressing creative power, it alone provides the necessary point of support in the aimless groping; and only it can uplift the spirit wearied by failure and urge it onwards to fresh efforts. A research worker who is not guided in his work by any hypothesis, however prudently and provisionally formed, renounces from the beginning a deep understanding of his own results. Whoever rejects the belief in the reality of the atom and the electron, or in the electro-magnetic nature of light waves, or in the identity of heat and motion, can most certainly never be convinced by a logical or empirical contradiction. But he must be careful how, from his point of view, he makes any advance in physical knowledge.

Indeed, belief alone is not enough. It is shown in the history of every science how easily it can lead to mistakes and deteriorate into narrow-mindedness and fanaticism. To remain a trust-worthy guide, it must be continually verified by logic and experience, and to this end the only ultimate aid is conscientious and often wearisome and self-denying effort. He is no scientist who is not at least competent and willing to do the lowliest work, if necessary, whether in the laboratory or in the library, in the open air or at the desk. It is in such severe surroundings that the view of the world is ripened and purified. The significance and meaning of such a process can only be realized by those who have experienced it personally.

Dynamical Laws and Statistical Laws

Conscientiousness and truth are as necessary in research in pure science as in practical life. The experimenter must not be blinded by the first results of a new intellectual discovery and must not neglect to prove conscientiously and thoroughly the results obtained in his researches. He must keep his mind fixed firmly on his original starting-point and the methods he employs. It may happen in a day that his hardly-won position is assailed and made untenable against severe criticism. Therefore, the experimenter cannot afford to close his eyes to a new discovery, obtained from another point of view, which will not fit in with his own ideas, nor must he treat it as unimportant, if not incorrect.

Such unforeseen and unexpected discoveries occur in all sciences, particularly when they are permeated with the spirit of youth. For every science, not even excluding mathematics, is to some extent the result of observation, whether the subject be natural or intellectual. The chief problem in every science is that of endeavouring to arrange and collate the numerous individual observations and details which present themselves, in order that they may become part of one comprehensive picture.

Now, considering the different subjects comprised in the various branches of science, their laws are by no means so diverse in their natures as might appear from a glance at the marked contrasts presented, for example, by questions in history and physics. At least it would be quite incorrect to look for a fundamental difference in the fact that in the domain of natural science a law must everywhere, without exception, be absolute, and the sequence of phenomena certain. In intellectual work, however, the pursuit of causal relations leads, here and there, always through something arbitrary and casual. On the one hand, every scientific thought involves the necessity of assuming a fixed absolute law raised above arbitrariness and chance to the highest level of the human intellect. On the other hand, even physics, the most exact of natural sciences, has frequently to deal with phenomena which cannot for the present be connected by any law, and which therefore may be considered accidental in a certain sense of the word.

Let us consider for a moment, as a special example, the behaviour of radio-active atoms according to the accepted disruption hypothesis of Rutherford and Soddy. How is it that a definite Uranium atom, after having remained completely unaltered and passive for untold millions of years, suddenly, in an immeasurably short space of time, without any determinable cause, explodes with a violence, compared with which our most powerful explosives are like toy pistols? It sends off fragments of itself with velocities of thousands of kilometres per second, and at the same time emits electro-magnetic rays of greater intensity than the hardest Röntgen rays, while another atom in its immediate neighbourhood, and to all appearances exactly similar, remains in a passive state for still more millions of years until finally it meets the same fate. In fact, all attempts to affect the course of radio-active phenomena by external means, such as raising or lowering of temperature, have ended in complete failure. It appears, therefore, at present hopeless even to guess at dynamical laws which would account for this. Yet the so-called theory of atomic disruption is of the greatest importance to physical research. It has co-ordinated, from the first, an almost embarrassing number of isolated facts, and has yielded many new results, some of which have been verified experimentally in a wonderful way, and others have stimulated new and important researches and discoveries.

How is this possible? How can practical laws be derived by considering phenomena the cause of which has, provisionally, to be left completely unexplained? Like the social sciences, physics has learnt to appreciate the great importance of a method completely different from the purely causal, and has applied it since the middle of last century with continually increasing success. This is the statistical method, and the newest advances in theoretical physics have been bound up with its development. Instead of seeking, without tangible results, the dynamical laws at present completely unknown to us, which govern a solitary occurrence, observations of a large number of isolated occurrences of a definite kind are collected and an average or mean value obtained. For the calculation of these mean values, certain empirical rules are available, according to the special circumstances of the case. These rules permit the prediction of future occurrences, not with absolute certainty, but with a probability which is often practically equivalent to certainty. This will not be true in all details, but only on the average, and that is usually what is wanted in applications.

Though a method which is fundamentally an expedient appears unsuited and unsympathetic to the scientific needs of many workers, who desire principally an elucidation of causal relations, yet it has become absolutely indispensable in practical physics. A renunciation of it would involve the abandonment of the most important of the more recent advances of physical science. It must also be borne in mind that physics, in the exact sense, does not deal with quantities that are absolutely determined; for every number obtained by physical measurements is liable to a certain possible error. Anyone who would only admit actual, definite numbers and not at the same time a possible error, would have to abandon the use of measurements and consequently all inductive knowledge.

It is sufficiently evident from the above that, in order to understand the characteristics of any science, it is of the utmost importance to differentiate carefully and fundamentally between the two classes of laws: the *dynamical*, strictly causal; and the solely *statistical*. I wish to compare and contrast these laws.

We will consider a few observations from everyday life. Let us take two open vertical glass tubes, connect the lower ends with rubber tubing, and pour into one of the tubes a quantity of a heavy liquid, such as mercury. The liquid will flow through the rubber tubing into the second tube, until the level of the surfaces in the two tubes is the same. This condition of equilibrium always returns after any disturbance. If, for example, one tube is suddenly raised, so that the mercury is for an instant raised with it, and consequently is at a higher level in that tube, it will immediately fall again until the surfaces are at the same height again in both tubes. This is the well-known principle on which every syphon action is based.

Let us take another example. We take a piece of iron, heated to a high temperature, and throw it into a vessel of cold water. The heat of the iron will be communicated to the water until complete equality of temperature is obtained. This is the so-called thermal equilibrium which will obtain after every disturbance.

There exists a certain analogy between the two examples. In each case, a certain difference brings about the variation, in the one case a difference of level and in the other a difference of temperature, and equilibrium is restored when the difference vanishes. Temperature is, therefore, sometimes referred to as level of heat. It can be said that, in the first case, the energy of gravitation, in the second case the energy of heat, flows from the higher level to the lower until the levels are the same.

It is no wonder that this analogy of a directing of energy has been explained as the action of a great general "principle of chance." This directing, though with the best intentions, has led to hasty generalizations. The principle makes each change in Nature an exchange of energy, and consider the different forms of energy as independent and of equal value. To each form of energy corresponds a factor of intensity, to gravitation height, to heat temperature, and the difference of these factors will determine the workings of chance. The confidence with which the general validity of this theorem was proclaimed is due to its simplicity and it was inevitable that it should appear early in popular expositions and elementary text-books.

Actually, the analogy between the two examples is only superficial, and the laws governing them are very widely separated. For, as all our experiences permit us confidently to assert, the first example obeys a dynamical law, the second a statistical one. Or, in other words, that liquid flows from a higher to a lower level is necessary, but that heat flows from a place of higher temperature to one of lower temperature is only probable.

It must be understood that such an assertion, which appears at first sight strange and almost paradoxical, requires to be supported by an enormous number of examples. I will endeavour to outline the most important of these and at the same time make clear the difference between dynamical and statistical laws. In the first place, that it is *necessary* for the heavy liquid to sink can easily be proved to be a consequence of the principle of conservation of energy. For if the liquid at the higher level rose to a level still higher without any external agency, and the liquid at the lower level sank still further, energy would be created out of nothing, which is contrary to the principle. The second case is somewhat different. Heat could very well flow from the cold water to the hot iron without violating the principle of conservation of energy; for, since heat is itself a form of energy, this principle only requires that the quantity of heat given up by the water is equal to that absorbed by the iron.

But the two operations show certain characteristic differences to the unbiased observer. The falling liquid moves faster the further it sinks. When the levels of the liquid are the same, the liquid does not come to rest, but moves beyond the equilibrium position on account of its inertia, so that the liquid originally at the higher level is at a lower level. Now, the velocity will decrease and the liquid will come to rest gradually, and subsequently the same process is repeated in the reverse direction. If all loss of kinetic energy at the air boundaries and that due to friction at the walls of the tube could be eliminated, the liquid would oscillate backwards and forwards indefinitely about its position of equilibrium. Such a process is, therefore, called reversible.

It is quite otherwise with heat. The smaller the difference of temperature between the iron and the water, the slower is the transmission of heat from the one to the other, and calculation shows that an infinitely long time elapses before equality of temperature is attained. In other words, there is always a small difference of temperature, however much time has elapsed.

There can be no talk of oscillation of heat between the two bodies: the flow of heat is always in one direction, and therefore represents an irreversible process.

In all physical science there is no more fundamental difference than that between reversible and irreversible processes. The former include gravitation, mechanical and electrical oscillations, acoustic and electro-magnetic waves. They can all be grouped under one single dynamical law—the principle of least action—which embraces the principle of the conservation of energy. Irreversible processes include conduction of heat and electricity, friction, diffusion and all chemical reactions, in so far as they take place with noticeable velocity. To cover these, R. Clausius derived his second law of thermo-dynamics, so exceptionally useful in physics and chemistry. The significance of this law is that it ascribes a direction to each irreversible process. But it was L. Boltzmann who, by the introduction of the atomic theory, explained the meaning of the second law and at the same time all irreversible processes, the peculiarities of which had presented insuperable difficulties of explanation by means of general dynamics.

According to the atomic theory, the heat energy of a body is simply the sum total of the extremely small, rapid, unregulated movements of its individual molecules. The temperature corresponds to the mean kinetic energy of the molecules, and the transmission of heat from a hot body to a cold body depends upon the fact that the kinetic energies of the molecules are meaned on account of the frequent collisions of the bodies. From this it must not be supposed that when two individual molecules collide, the one with the greater kinetic energy is slowed up and the other accelerated, for if, for example, a rapidly moving molecule of one system is struck obliquely by a slower moving molecule, its velocity must be still further increased, while that of the slower molecule is still further diminished. But, in general, unless the circumstances are very exceptional, the kinetic energies must mix to a certain extent, and this corresponds to an equalizing of the temperatures of the two bodies. All results deduced in this manner agree with observation, particularly in the case of gases.

However much discussed and however promising this atomic theory might appear, it was, until recently, regarded merely as a brilliant hypothesis, since it appeared to many far-sighted workers too risky to take the enormous step from the visible and directly controllable to the invisible sphere, from the macrocosm to the microcosm. In order that he should not imperil the acceptance of his observations and calculations, Boltzmann himself did not over-emphasize them. He laid stress on the view that the atomic hypothesis was a mere representation of what took place. To-day we may go further towards comparing the reality with the picture, in so far as it has any meaning at all, from the point of view of the philosopher. For to-day we have a series of experiments which invest the atomic hypothesis with the same degree of certainty as is possessed by the mechanical theory of sound, or the electro-magnetic theory of light and heat radiations.

According to the theory of chance, inadequately outlined above, the condition of a stationary fluid of uniform temperature must be absolutely invariable; for if no difference of intensity of any sort exists in the fluid, there can be no cause which will bring about any variation. The state of a fluid can be made visible by introducing into a transparent liquid, such as water, a number of minute particles or drops of another liquid, such as gummastic or gamboge. I do not think that anyone who has observed such a preparation through a properly illuminated microscope, can ever forget his first view of the play presented to him. It is a glance into a new world. Instead of the complete tranquillity he expected, he sees an extraordinarily lively, gay dance of the small floating particles, in which the smallest behave in the most erratic manner: no trace of any friction in the fluid can be seen; if a particle once comes to rest, another starts the game. One is involuntarily reminded of the frenzied activity of an ant-hill which has been disturbed. But whereas the angry insects gradually calm down and lose their activity towards dusk, the particles under the microscope never show the least signs of fatigue, while the temperature of the liquid remains unaltered—an actual case of perpetual motion, in the most literal sense of this much-used expression.

The phenomenon described was discovered in the year 1827 by Brown, the English botanist, but it had been deduced by the French physicist Gouy, twenty-five years earlier, from the movements of molecules in a heated fluid. These molecules, themselves invisible, continually collide with particles floating around them (which are visible in a microscope) and are impelled along irregular paths. The final theoretical proof of the correctness of this explanation was first given quite recently, when Einstein and Smoluchowski, obtained statistical laws governing the distribution of density, the velocities, the mean free paths, and even the rotations of the microscopic particles, and these laws were most strikingly confirmed quantitatively in all details, particularly through the experimental work of Jean Perrin.

There can be no doubt now, in the mind of the physicist who has associated himself with inductive methods, that matter is constituted of atoms, heat is movement of molecules, and conduction of heat, like all other irreversible phenomena, obeys, not dynamical, but statistical laws, namely, the laws of probability. Indeed, it is difficult to make even an approximate estimate of the probability that heat will travel in the contrary direction, i.e. from the cold water to the hot iron. If one draws one letter after another at random from a sack filled with letters, and sets them out in a row in the order in which they are drawn, there is always a *possibility* that complete words may be formed, even that they will form a poem by Goethe. Or if a hundred throws are made with a die, no one will dispute the possibility that six will turn up each time without exception, since the result of each throw is independent of the previous one. Should this occur in practice, there is no doubt that everyone would say that there was something wrong, perhaps the die was not quite symmetrical, and no rational person would deny the weight of this observation. For the probability that so exceptional an occurrence should take place under normal circumstances is extremely minute. Yet this is enormously great compared with the probability that heat will flow from a cold to a hot body. We need only consider that in the case of the die, we are dealing with six numbers, consequently with six different cases, in the case of the letters, with twenty-six, but in the case of the molecules with many millions in the smallest visible space, and moving with extremely diverse velocities. Thus from the standpoint of practical physics, there is certainly no ground to believe the possibility of a deviation from the general truth of the laws governing radiation of heat.

This is certainly not the case with the theory. For it is clear to everybody that there must be an unfathomable gulf between a probability, however small, and an absolute impossibility. This can be demonstrated in particular circumstances. One need only throw the die sufficiently often in order, with greater probability, to expect a hundred consecutive sixes, and one need only persevere sufficiently long at the letter game to obtain a Faust monologue. Still, it is as well that we do not depend solely on these methods, for neither the age of a man, nor probably that of mankind, would be long enough.

Whatever the application to physics involves, it is necessary to consider very seriously such infinitesimal probabilities under certain conditions. If a powder magazine were to explode without any determinable cause, the occurrence would not be ignored. The so-called self-ignition is to be regarded as caused by a very improbable accumulation of dangerous impacts of chemically reacting molecules; the laws governing these molecules can only be arrived at statistically. It is obvious that in an exact science such words as *certain* and *sure* must be used with great caution, and the importance of the laws of observation must be very moderately assessed. Thus, when considering the laws of physics, or, indeed, any observed law, either dynamical or statistical, we are compelled by theory and experiment alike to make a fundamental difference between necessity and probability. This duality, which has been brought into all physical laws by the introduction of statistical methods, will appear unsatisfactory to many. Accordingly, when it appeared unsuitable, efforts were made to set it aside by denying absolute certainty and impossibility, and substituting great and small degrees of probability respectively. If there were no dynamical laws in Nature, but only statistical, the conception of absolute necessity would have no place in physics. Such a view must very soon prove to be a mistake as dangerous as it is short-sighted, apart from the fact that all reversible processes, without exception, are governed by dynamical laws, and that we have no reasons for discarding these laws. Physics can no more do without the hypothesis of absolute laws than can any other natural science or human study, for without it the essential foundations of deductions from statistics would be removed, and it is these deductions that we are considering.

Yet one considers that the theorems of the calculus of probability are not only capable of, but also require, a strict exposition and rigid proof, and therefore it has always particularly attracted prominent mathematicians. If the probability that a certain event is succeeded by a certain other event is $\frac{1}{2}$, then, it can be said that nothing is known of the occurrence of the second event, except that it will follow in just 50 per cent. of the cases when the first event occurred, and that this percentage is more nearly obtained the greater the number of cases that are considered. In addition, the calculus of probability furnishes an exact estimate of the deviation from the mean, which is to be expected when the number of observed cases is smaller, i.e. of the so-called dispersion. If the observations are in contradiction to the calculated magnitude of the dispersion, it may be safely concluded that an erroneous assumption was made in the premises, a so-called systematic error.

To support such far-reaching assertions, very extensive presuppositions are naturally essential, and it will be understood that in physics the exact calculation of probabilities is only possible when purely dynamical laws can be assumed to hold in the simplest occurrences, i.e. in the smallest microcosm. Should these laws contradict a single observation through our fallibility, the hypothesis of their absolute immutability furnishes a necessary firm foundation for the structure of statistics.

It appears from these remarks that the duality between statistical and dynamical laws is intimately associated with the duality between macrocosm and microcosm, and this we must regard as a fact substantiated by experiment. Whether satisfactory or not, facts cannot be created by theories, and there is no alternative but to concede their appointed places to dynamical as well as to statistical laws in the whole system of physical theories.

Thus dynamics and statistics cannot be regarded as inter-related. For, whereas a dynamical law completely satisfies the causal requirements and is therefore of a simple character, every statistical law is built up, and it cannot in any way be looked on as definitive, since it always involves the problem of reduction to its simple dynamical elements. The solution of such problems is one of the chief tasks of progressive science. This is as much the case in chemistry as in the physical theories of matter and in electricity. Meteorology may also be mentioned in this connection, for the work of V. Bjerknes provides a scheme of great magnitude to reduce all meteorological statistics to their simpler elements, namely, to physical laws. Whether the attempt leads to practical results or not, it must be made at some time, since the essence of all statistics is that while it often has the first, it never has the last, word.

As the principle of conservation of energy or the first law of thermo-dynamics occupies the first place among the dynamical laws of physics, so the second law of thermo-dynamics holds a corresponding place among the statistical laws. Although this theorem is a probability theorem, and, in consequence, one often speaks of limits to its validity, it can be expressed in an exact and generally valid form. It might be expressed somewhat as follows: All physical and chemical changes of state proceed, on the average, towards states of greater probability. Of all the states that can be assumed by a system of bodies, the most probable is that in which all the bodies have the same temperature. On this ground only is based the law that heat conduction always, *on the average*, tends towards an equalization of temperatures, and also from the higher to the lower temperature. The second law will only allow us to deduce anything with certainty from a *single* observation if we are certain beforehand that the course of the operation in question is not markedly different from the mean course deduced from a large number of operations in which the initial conditions were the same. To make sure that this condition is satisfied, it is, theoretically, sufficient to introduce the so-called hypothesis of elementary disorder. Experimentally, the only method is to repeat the particular observation many times, or to have it done by different observers, working independently of one another. Such a repetition of a definite experiment, or the arranging of a whole series of experiments, is actually what is done in practical physics. For, in order to eliminate unavoidable errors of observation, no physicist will limit himself to the results of a single experiment.

The second law of thermo-dynamics has nothing to do with energy directly. A good example of a process which need not be accompanied by a transformation of energy is diffusion. Diffusion happens solely because a uniform mixing of two different substances is more probable than a non-uniform mixing. This can, indeed, be subordinated to the conception of energy, by introducing, for this special purpose, the idea of free energy, which permits of a convenient exposition and in many cases simplifies the representation. The method, however, is indirect in so far as free energy can only be understood from its relation to probability.

Let us, in conclusion, pause awhile after this rapid survey to consider the laws of phenomena in the intellectual sphere. To a great extent, we find quite similar relations, except that causality is completely eclipsed by probability, the microcosm by the macrocosm. Yet here in all questions extending to the highest problems of intellect and morality, the assumption of absolute determinism is a necessary basis for every scientific investigation. Care must be taken that the normal course of the phenomenon examined is not disturbed by the examination. This is equally true in natural science, but not usually emphasized on account of its being almost self-evident. When a physicist wants to take the temperature of a body, he does not use a thermometer the introduction of which would alter the temperature of the body. From this point of view the possibility of a completely objective scientific investigation into psychological phenomena only extends to the critical examination of personalities other than the observer, so long as they are independent of the observer. In so far as it is completely effaced from the mind of the investigator, it also extends to the past, but not to the present, nor to the future, which must always be attained through the present. Thought and research are themselves psychological phenomena in man, and if the object of the investigation is identical with the investigator, he must change continually as his knowledge advances.

It is, therefore, quite useless to treat exhaustively the phenomena of the future from the standpoint of determinism, and with it to wish to fix the conception of moral freedom. Self-determination is given to us by our consciousness and is not limited by any causal law, and he who considers it logically irreconcilable with absolute determinism in all spheres of philosophy, makes a great mistake of the same nature as that made by the physicist already referred to, who does not take adequate precautions to eliminate errors in his observations, or a mistake such as a physiologist would make if he examined himself in order to study the functioning of a muscle in anatomy.

Science thus fixes for itself its own inviolable boundaries. But man, with his unlimited impulses, cannot be satisfied with this limitation. He must overstep it, since he needs an answer to the most important, and constantly-repeated question of his life: What am I to do?—And a complete answer to this question is not furnished by determinism, not by causality, especially not by pure science, but only by his moral sense, by his character, by his outlook on the world. Conscientiousness and truth are the ideals that will lead him along the true path in life as in science. They will guarantee him, not necessarily brilliant results, but the highest good of humanity, namely, inward peace and true freedom.

The Principle of Least Action

As long as physical science exists, the highest goal to which it aspires is the solution of the problem of embracing all natural phenomena, observed and still to be observed, in one simple principle which will allow all past and, especially, future occurrences to be calculated. It follows from the nature of things, that this object neither has been, nor ever will be, completely attained. It is, however, possible to approach it nearer and nearer, and the history of theoretical physics shows that already an extensive series of important results can be obtained, which indicates clearly that the ideal problem is not purely Utopian, but that it is eminently practicable. Therefore, from a practical point of view, the ultimate object of research must be borne in mind.

Among the more or less general laws, the discovery of which characterize the development of physical science during the last century, the principle of Least Action is at present certainly one which, by its form and comprehensiveness, may be said to have approached most closely to the ideal aim of theoretical inquiry. Its significance, properly understood, extends, not only to mechanical processes, but also to thermal and electro-dynamic problems. In all the branches of science to which it applies, it gives, not only an explanation of certain characteristics of phenomena at present encountered, but furnishes rules whereby their variations with time and space can be completely determined. It provides the answers to all questions relating to them, provided only that the necessary constants are known and the underlying external conditions appropriately chosen.

This central position attained by the principle of least action is, however, not even to-day quite undisputed; for a long time the principle of conservation of energy has been a keen competitor. The latter governs in a similar manner the entire range of physics and certainly possesses the advantage of being more easily explained. It is, therefore, advisable to examine briefly the relative positions of these two principles.

The principle of conservation of energy can be derived from the principle of least action and is consequently contained in it. The converse is, however, not true. Accordingly, the former is the more particular, and the latter the more general principle. As an illustrative example, let us consider the motion of a free particle under no forces. According to the principle of conservation of energy, such a particle moves with constant velocity, but nothing is said concerning the direction of the velocity, since kinetic energy does not depend on direction. The path of the particle could, for example, be rectilinear or curvilinear. On the other hand, the principle of least action demands, as we shall show in detail below, that the particle must move in a straight line.

Now in this simple example an attempt could be made to extend the principle of conservation of energy by making certain simple assumptions, such as that not only the total kinetic energy of the moving particle remain constant, but that also the component of the energy along a certain given direction in space be constant. Such an extension would be foreign to the principle of energy and would be difficult to apply to more general problems. In the case of the spherical pendulum, that is a heavy particle in motion on a fixed sphere, this principle could only furnish the following solution. During the upward motion, the kinetic energy decreases in a certain manner, and increases during the downward motion. The path of the particle cannot however be determined, whereas the principle of least action completely solves all questions bearing on the motion.

The reason for the difference in the results derived from the two principles lies in the fact that when applied to any problem, the principle of conservation of energy furnishes one equation only, while it is necessary to obtain as many equations as there are independent variables in order to determine the motion completely. Thus in the case of the free particle three equations are needed, and in that of the spherical pendulum two. Now, the principle of least action furnishes, in every case, as many equations as there are variables. Moreover, it enables several equations to be embraced in one formula, as it is a variations principle, as opposed to the principle of energy.

From the infinite number of virtual motions imaginable under the given conditions, it indicates a quite definite motion by means of a simple criterion, and shows that this is the actual motion. The criterion is that in the transition from the actual motion to an arbitrary motion infinitesimally close to it, or more accurately, for every infinitesimally small variation of the given motion consistent with the given conditions, a certain function characteristic of the variation vanishes. By this means one equation is derived from every independent variable, as in the case of a maximum or minimum problem.

Now, it must be understood that the principle of least action only attains a definite significance when we have given to us the prescribed conditions with which the virtual motions must be consistent, as well as the characteristic functions which vanish for every arbitrary variation of the actual motion. The problem of determining the correct conditions always forms the essential difficulty in formulating the principle of least action. However, it must be obvious that the idea of combining into one variations principle a number of equations, necessary for defining the motion of any complicated mechanical system is in itself of great importance and represents an appreciable advance in theoretical research.

In this connection mention may certainly be made of Leibniz's theorem, which sets forth fundamentally that of all the worlds that may be created, the actual world is that which contains, besides the unavoidable evil, the maximum good. This theorem is none other than a variations principle, and is, indeed, of the same form as the later principle of least action. The unavoidable combination of good and evil corresponds to the given conditions, and it is clear that all the characteristics of the actual world may be derived from the theorem, even to the details, provided that, on the one hand the standard for the quantity of good, and on the other hand the given conditions, be rigidly defined along mathematical lines—the second is just as important as the first. Before, however, we can hope to derive important results from the principle we must advance still further. First of all, the characteristic quantity, which vanishes in the case of the actual motion, must be investigated and understood. We may proceed from two different points of view. According to one, the characteristic function is referred to an isolated time point or to an infinitesimal time element; according to the other, it is referred to a finite time interval during the motion. We arrive at two different classes of variations principles according as to whether we decide to adopt the one or the other standpoint.

To the first class belong Bernoulli's Principle of Virtual Displacements, d'Alembert's Principle of Resistance, Gauss's Principle of Least Constraint, and Hertz's Principle of the Shortest Path. All these principles may be considered as differential principles in so far as they apply the characteristic criterion to a property of the motion which is referred to an isolated movement or to a small element of time. In the case of mechanical systems any one of them is completely equivalent to any other and to Newton's Laws of Motion. But they all suffer the disadvantage that it is only for mechanical systems that they have any significance, and their exposition renders it necessary to introduce special point co-ordinates for the mass systems under consideration, and varies with the choice of co-ordinates. The exposition is usually comparatively complicated.

The inconvenience of such mechanical systems of co-ordinates may be overcome if, as a matter of course, the variations principle be considered as an Integration Principle by referring it to a finite time interval. Then, of all virtual motions, the actual motion is that defined by the property that for any permissible variation from it a certain time integral vanishes. In the most important cases, this condition can be expressed as follows: For the actual motion, a certain time integral, which may be called "The Magnitude of the Action"* or the "Action"† of the motion, is less than that of any other motion consistent with the prescribed conditions. Thus, according to Leibniz, the action of

---

* Wirkungsgrösse.   † Aktion.

any single material particle is equal to the time integral of the kinetic energy, or, in other words, the time integral of the velocity.

In this manner, the principle of least action can be applied without reference to any special system of co-ordinates, and without pre-supposing any mechanical phenomenon, since only time and energy appear in the expressions. A special feature appears through the introduction of the time integral, and the presence of this feature has always been, and is even now, considered by many physicists and philosophers to be a criticism to be levelled against the principle of least action as against every other variations principle. Thus, by referring to a finite time interval, the motion at any instant is investigated with the help of a later motion, and present events are in a certain manner made dependent upon later events, and the principle acquires a teleological character. When dealing with the principle of causality, it must be possible to understand and derive all the characteristics of a motion from previous circumstances, without reference to anything that may happen later. This appears not only feasible but a direct logical consequence. On the other hand, when seeking the most lucid relations in the system of natural laws, such aids as reference to later events will be considered permissible in the interests of the desired harmony. These may not be directly essential for the complete exposition of natural phenomena, but they might, perhaps, be more convenient to handle, or more easily interpreted. I would remind you that in order to retain the symmetry of equations in mathematical physics, it oftens happens that the quantities to be determined are not expressed in terms of the independent variables themselves. On the contrary, one or more superfluous variables are often introduced in order to utilize the great practical advantages of symmetry.

Since Galileo's time, physics has achieved its greatest successes by rejecting all teleological methods. It is justified, therefore, in definitely opposing all attempts at introducing teleological points of view into the law of causality. However, though the introduction of a finite time integral is unnecessary for formulating the laws of mechanics, yet the integral principle should not, as a matter of course, be rejected. The question of its correctness has nothing to do with teleology. It is far more a practical question, and may be thus expressed: does the exposition of the laws of Nature by means of the integral principle accomplish more for the purposes of theoretical physics than other expositions? From the modern standpoint, the answer must be in the affirmative, on account of the fact that the integral principle, as already mentioned, is independent of any special co-ordinates. The modern principle of relativity provides, as we shall see later, a complete explanation, not only of the practical significance, but also of the need for introducing the finite time interval into the fundamental principles of mechanics.

In the exposition already given of the principle of least action, no account has been taken of the prescribed conditions of the virtual motions. These are, however, quite as important as the magnitude of the *action* itself, for the significance of the principle differs with the nature of the prescribed conditions. It is not only a question of how the selection is made, but also of the nature of the motions determined by the choice. This circumstance was at first overlooked, and many serious errors were thereby introduced. It was a long time before it was clearly explained, and the principle of least action correctly understood. If the principle be said to have been discovered at this time, the honour should be given to Lagrange. This, however, would be an injustice to other men who had prepared the way for Lagrange to bring the work later to a satisfactory completion. Of these, the first was Leibniz; indeed, he was the chief, according to a letter dated 1707, the original of which has been lost. Then came Maupertuis and Euler. It was chiefly Moreau de Maupertuis (appointed president of the Prussian Academy of Sciences (1746–1759) by Frederick the Great) who not only recognized the existence and significance of the principle, but used his influence in the scientific world and elsewhere to procure its acceptance. Maupertuis repeatedly announced in different forms, his principle of *Mitwelt*, and zealously defended it against what were often authoritative criticisms. The zeal with which he did this rose at times to fanaticism, and was quite disproportionate to the scientific value of the enunciation considered most suitable by him. It is impossible to reject the idea that his energetic adherence to really unattainable theses, arose not alone from scientific conviction, but at least equally from a firm intention to ensure for himself, at all costs, the prior claim to the discovery of what he regarded as his most important work. This is especially shown in the passionate attempts he made to dispute Leibniz's letter (already referred to) when it was produced by Professor Samuel König in 1751—attempts which almost led him to abuse the high position he occupied. Human weakness and vanity have hardly ever been more severely punished than in the case of the president of the Berlin Academy. His varying fortunes, which occasionally induced the great royal philosopher to interfere, have been repeatedly described in detail by historians and in technical literature by A. Mayer (1877), H. von Helmholtz (1887), E. du Bois-Reymond (1892), and H. Diels (1898). An account of the discussion, from the standpoint of the general development of mathematical science, is given in Cantor's "History of Mathematics," and its significance for the Berlin Academy is dealt with in Harnack's history of the Academy.

Maupertuis's exposition of the principle of least action asserted no more than "that the action applied to bring about all the changes occurring in Nature is always a minimum." Strictly, this formulation does not admit any conclusions to be drawn regarding the laws governing the changes, for as long as no statement of the conditions to be satisfied is made, no deductions can be made as to how the variations are balanced. Maupertuis had not the faculty of analytical criticism necessary to discern this want. The failure will be more easily understood when it is realized that Euler himself, a brilliant mathematician, did not succeed in producing a correct formulation of the principle, though he was assisted by many colleagues and friends.

Maupertuis's real service consisted in his search for a principle that would be, above all, a minimum principle. That was the real object of his investigation. To this end he made use of Fermat's principle of quickest arrival, although its bearing upon the principle of least action was very indirect and, at all events, unknown to the physics of his time. The interest in the principle of least action was fundamentally based upon the metaphysical idea that the rule of the Deity reveals itself in Nature. Therefore, every natural occurrence is founded on an intention which is directed to a certain end, and which indicates the most direct way and the most suitable methods towards attaining this end.

How inadequate, and even misleading, teleological methods can be, is best realized from the fact that, from a general point of view, the principle of least action is not, strictly speaking, a minimum principle at all. Thus, the statement that the path of a particle, free to move without friction on a sphere, is the shortest line joining its initial and final positions, is not true if the path is longer than the semi-circumference of a great circle on the sphere. Beyond the semi-circumference, therefore, Divine foresight cannot operate. Still more striking is the fact that when considering non-holonomous systems, the virtual motions bear no relation to the possible motions, and thus the minimum condition loses all its significance.

In spite of all this, however, it must be borne in mind that the strong conviction of the existence of a close relation between natural laws and a higher will has provided the basis for the discovery of the principle of least action. Provided, of course, that such a belief is not confined within too narrow limits, it certainly does not admit of proof, but, on the other hand, it can never be disproved, for then one could ultimately ascribe any contradiction to an incomplete formulation.

J. L. Lagrange was the first to express correctly the principle of least action (1760). Thus—of all the motions that may bring a system of material particles from a certain initial position to a given final position (the total energy remaining constant), the actual motion is that for which the action is a minimum. The virtual motions must, therefore, satisfy the principle of energy. They may, on the other hand, take any arbitrary time. According to this conception, the path of a particle is that along which it will reach its final position in the shortest time, if it move with constant velocity, and if frictional forces be absent. Thus, the path is the line of shortest length, that is, for a free particle a straight line.

Later, C. G. J. Jacobi and W. R. Hamilton showed that the principle admitted of other representations. Hamilton's exposition was of great importance from the standpoint of future developments. According to him, the total energy of the virtual motions to be compared need not remain constant, but the motions must take place in the same time. Then, however, the action which is a minimum for the actual motion, must not be expressed by Maupertuis's time integral of the kinetic energy, but by the time integral of the difference of kinetic and potential energies. Applying this method to the above example of a particle in motion when not affected by frictional forces, the principle shows that of all the possible curves, the actual path is that along which the particle reaches its final position in a given time with the least velocity, again the line of shortest length.

In a characteristic way, the principle of least action did not at first exercise an appreciable effect on the advance of science, even after Lagrange had completely established it as a part of mechanics. It was considered more as an interesting mathematical curiosity and an unnecessary corollary to Newton's laws of motion. Even in 1837 Poisson could only call it "a useless rule." It was in the investigations of Thomson and Tait, G. Kirchhoff, C. Neumann, L. Boltzmann and others that the principle first proved itself to be an excellent method for solving problems in hydro-dynamics and elasticity. While the usual methods of mechanics sometimes worked with difficulty and at other times refused to work, a revolution was in the making— the value of the principle began to be realized. In 1867 Thomson and Tait wrote concerning it, "Maupertuis's celebrated principle of least action has been, even up to the present time, regarded rather as a curious and somewhat perplexing property of motion than as a useful guide in kinetic investigations. We are strongly impressed with the conviction that a much more profound significance will be attached to it, not only in abstract dynamics, but in the theory of the several branches of phsyical science now beginning to receive dynamic explanation."

It was also shown that when applying the principle, especially when defining the prescribed conditions of the virtual motions, particular care was necessary if errors were to be avoided. For example, when considering the irrotational motion of an inviscid fluid round a solid body, it is, in general, not sufficient to assume the initial and final positions of the body given, the initial and final positions of the fluid elements must also be given. H. Hertz made an error of another type, in the introduction to his mechanics, when he applied the principle of least action to investigate the motion of a sphere rolling on a horizontal plane, and assumed for the virtual displacements certain conditions not allowed when dealing with non-holonomous systems. O. Hölder and A. Voss did much to make this problem clear.

The fundamental significance of the principle of least action, as a general principle, was first understood when it was realized that it could be applied to systems whose mechanism was either entirely unknown, or so complicated that they could not be considered by means of ordinary systems of co-ordinates. After L. Boltzmann and later R. Clausius had perceived the close relation between the principle and the second law of thermo-dynamics, H. von Helmholtz gave, for the first time, a complete and systematic summary of such applications of the principle as were possible at the time to the three great branches of physics—mechanics, electro-dynamics, and thermo-dynamics. This was a surprising achievement in view of the comprehensiveness of the range covered.

For his calculations, Helmholtz chose Hamilton's form of the principle as being the most convenient, and made some extensions of a formal nature. He used the term kinetic potential to denote the quantity the time integral of which was what Hamilton called the action. He thus retained the hypothesis that the principle was fundamentally a mechanical one. This limitation, however, was somewhat of a retrogression, since it was not necessary to consider the mechanical constitution of several of his systems, such as galvanic streams and magnets. On the other hand, Helmholtz accomplished the deciding act, in that he did not derive his kinetic potential from the difference of the kinetic and potential energies as had hitherto been done, but he set forth the kinetic potential as the primary quantity, and thus determined the magnitude of the energy and all the remaining laws of motion.

The chief consequence of this new method of considering the question was an immediate generalization of some importance. The kinetic potential is not only the analytical form of the energy, but gives also its magnitude, which varies according to the choice of the independent variable. For example, some of the equations of motion can be used in order to reduce the number of independent variables. The variables eliminated have then disappeared from the principle entirely and may be said to correspond to the concealed motion. In each such case, the kinetic potential assumes a different magnitude, and thus are explained, for example, the expressions derived for the potential in thermo-dynamics which differ with the choice of the independent variable. Helmholtz showed how these different expressions are inter-related, and follow from one another; he also showed that the kinetic potential can assume a form in which it appears no longer as the difference between the kinetic and potential energies. This result demonstrated at once the universality of the principle, for outside the range of mechanics the distinction between kinetic and potential energies is no longer possible, and, therefore, the possibility of deriving the kinetic potential uniquely from the energy disappears, while in each case the converse is simple.

Although it had been possible for Helmholtz to hold fast to the assumption (at least in principle) that all physical phenomena can ultimately be reduced to the motions of simple particles, considerable doubt has since been thrown on the validity of the assumption, at least as far as electro-dynamics is concerned. There is no doubt, however, from all the results hitherto obtained, that the principle of least action has been proved to be applicable and useful in physics outside the range of mechanics, especially in the electro-dynamics of absolute vacuum. Without making use of any mechanical hypothesis, J. Larmor (1900), H. Schwarzschild (1903), and others have derived the fundamental equations of electro-dynamics and electron theory from the Hamiltonian principle.

Thus the development of the principle of least action has

followed along similar lines to that of the principle of conservation of energy. The latter was also originally regarded as a
mechanical principle, indeed its general validity was directly
considered, for a long time, to be a suitable basis of the mechanical view of Nature. To-day, the mechanical conception of
Nature has lost ground, while there has never been any occasion
to doubt the universality of the principle of energy. Anyone
desiring to regard the principle of least action as mechanical
would to-day have to apologize for doing so.

The most brilliant achievement of the principle of least action
is shown by the fact that Einstein's theory of relativity, which
has robbed so many theorems of their universality, has not disproved it, but has shown that it occupies the highest position
among physical laws. The reason for this is that Hamilton's
"Action" (not Maupertuis's) is an invariant with respect to all
Lorentz transformations, that is, it is independent of the system
of reference of the observers. This fundamental characteristic
gives a far-reaching explanation of the striking circumstance
(and at first sight unfortunate) described on page 72, namely,
that "action" refers to a time interval, and not to an instant of
time. In the theory of relativity time plays a part analogous to
space. According to the theory of relativity the problem of
determining the state of a system of bodies in different positions
at any time from the state in different positions at any given
time is exactly similar to the problem of determining the state at
different times in all positions from the state of the system at
different times in any given position. Though the first problem
is usually regarded as the real problem of physics, yet, strictly
speaking, there is in it a certain arbitrariness and unreal
limitation, which only finds a historical explanation in the fact
that its solution is, in the majority of cases, of greater use to
humanity than that of the second. Now, just as the calculation
of the action of a system of bodies necessitates an integration
over the space occupied by the bodies, the action must contain
also a time integral in order that no priority is given to space
over time, for space and time together constitute the universe to
which the action relates.

As in the case of the principle of least action, the principle of the conservation of energy has also a special position in the theory of relativity. Energy is, however, not an invariant with reference to Lorentz transformations any more than it was earlier with respect to Galileo's transformations. For in energy time plays the more important part. The corresponding principle into which space enters is the principle of conservation of momentum. The principle of least action stands superior to both, even when considered together, and it appears to govern all the reversible processes of Nature. Nevertheless, it offers no explanation for irreversibility, since according to it, all phenomena can proceed backwards or forwards in any direction in space and time. That is why the problem of irreversibility has not been considered in this paper.

The Relation between Physical Theories

It is well known that no science develops systematically from one single starting-point according to a definite preconceived plan, but that its development depends upon practical considerations and proceeds more or less simultaneously along different lines, corresponding to the many ways of looking at the problems, and to the time and views of the investigator. A science is developed in many places, along many different lines and in many periods of time. It frequently happens that theories are found to be inter-related which were started from essentially different view-points; theories, when extended and completed, turn out to be similar and begin to influence one another, appearing helpful or inimical to each other according to circumstances. This constitutes the characteristic difference between mathematics and an experimental science. In the former, two different theories are never mutually contradictory, as each is correct—in mathematics it is not possible to contrast theories but only methods; for example, as a matter of course it is impossible for an algebraic theory to contradict a geometrical one, though algebra and geometry have developed quite independently of one another. On the other hand, in physics as a practical science it has frequently happened and still does happen that two theories, developed independently of one another, come into conflict when extended and must be mutually modified to remain compatible. In this mutual adjustment lies the germ of their further development into a complete unity, since the chief purpose of each science is, and always will be, the unifying of all its great theories into one which will embrace all the problems of that science and afford a solution to all of them. From this point of view it can be said that the science which is nearest its goal is that one which has combined the greatest number of its theories. The history of physics offers numerous instructive examples of this process of mutual adjustment and coalescence. In this short sketch, the historical development will be considered only as far as is necessary, and attention will be drawn chiefly to the present position of the physical theories.

Even to-day, physics can be divided into three essentially different groups of theories: *mechanics*, including elasticity, hydro-dynamics, and acoustics, then *electro-dynamics*, with magnetism and optics, and *thermo-dynamics*. Each of these three groups of theories has retained a certain degree of independence, though there are to-day a large number of points of contact between them, either supplementing or clashing with each other. Thanks to the rapid advance of experimental science, the number of these points of contact is continually increasing.

The oldest and first to be developed of the theories of physics is mechanics which, therefore, claimed the attention of all the early thinkers in the subject, and should, according to many physicists, still have the prior claim. It was founded by Galileo and Newton, and brought to its final form by Euler and Lagrange. This branch of physics forms a complete picture, leaving nothing to be desired in rounding off and filling in, and it can emulate a mathematical theory in strictness. It is just in this finality, peculiar to classical mechanics, that lies the impossibility of further expansion and development, demanded by the general problem of physics, which has to explain a large number of observations in addition to the phenomena of motion. In fact, the impulse for an extension of mechanics had to come from without, and it came from the electro-dynamical theory. It is of peculiar interest to note how this theory, at first partly dependent on the older, more mature, mechanics, gradually drifted away from it, developed along independent lines and finally grew to such an extent that it could exert an almost revolutionary influence on the classical mechanics.

Though, as we have said, electro-dynamics developed entirely under the influence of mechanics, its development on the Continent was quite different from that in Britain. In Germany, the method was indicated by Gauss, a mathematician and astronomer, who brought electrical phenomena into agreement with Newton's law of gravitation and, therefore, sought the fundamental theorem of electricity in a generalization of Newton's law of distance effect. According to him, the element in electro-dynamics was a quantity of electricity of electric mass which is a counterpart of ponderable mass, and the general fundamental theorem in electricity differs from the Newtonian law in that the force exerted by two electrical elements on one another, besides depending on the magnitudes of, and distance between, the elements, involves their signs and velocities. Definite forms of such a fundamental electrical law were advanced by W. Weber, B. Riemann, and R. Clausius. The development of electro-dynamics followed quite different lines in England, where Faraday impressed upon it the mark of his genius, in that he studied electrical phenomena directly without being in any way influenced by mathematics or astronomy, and so brought them into line with elasticity. He considered as the unit not the electric charge but the electric lines of force that run from one charged body to another, and which correspond to mechanical stresses in the intervening medium; a direct distance effect is thus entirely excluded. Subsequently, Maxwell expressed Faraday's hypothesis in mathematical form, employing mechanical conceptions totally different from those used by Gauss. This theory proved itself equal, and later superior, to all distance theories, and the differences came in that part most intractable to classical mechanics, that dealing with phenomena in a pure vacuum.

It is true that an absolute vacuum cannot be obtained in Nature, but many experiments, such as Fizeau's optical measurements, have shown that electrical phenomena, particularly optical ones, in a very rarefied gas are completely independent of the nature of the gas, so that in practice an absolute vacuum can be said to exist for the purposes of physics. Here the distance theories fail, unless the complicated assumption, foreign to the idea of distance effect, is made that the pure vacuum can be polarized, whereas Maxwell's theory assumes the most simple and clear form for this simplest of all media.

Now, while Maxwell's equations obtained their signal

triumph on this question, as time went on greater difficulties arose in connection with the numerous investigations into the mechanical foundations of these equations, which presuppose the existence of a material substratum, viz. light-ether. To-day, it is generally accepted that an absolutely rigid mechanical theory of ether consistent with Maxwell's simple equations cannot be countenanced and this makes a gap between classical mechanics and electro-dynamics that cannot be bridged. It only remains to define exactly the range of applicability of the two theories or to modify one of them. The former soon proved to be impracticable, for mechanics and electro-dynamics overlapped on the question of motion of electrons and the manner of settling the controversy was shown for the first time by the discovery of deviations from the laws of classical mechanics, the deviations finding expression in the variable inert mass of the electron. Einstein's theory of relativity contains a simple, complete solution of the problem of reconciling mechanics and electro-dynamics in a quite general way, in so far as it retains the practical essentials of the classical theory and still fulfils the demands of electro-dynamic calculations. The modification in mechanics due to the principle of relativity is the introduction of a new universal constant, entirely foreign to classical mechanics, namely, the velocity of light in an absolute vacuum.

While mechanics and electro-dynamics are preparing to combine under the banner of the principle of relativity into a unified theory, which I shall call *dynamics* in the following, there remains the last great problem of theoretical physics, viz. the correlation of dynamics with the theory of heat; a problem that has already been taken in hand successfully but which presents enormously greater difficulties than have previously been encountered, since the nature of the laws which govern dynamics is, in some respects, totally different from that of the laws governing the theory of heat. It is chiefly the ever-recurring characteristics of the irreversibility of all heat phenomena: a characteristic not capable of dynamical explanation. While, in fact, all thermal and chemical actions proceed in one direction only, the sign of the time variable plays no part in the equations of dynamics, i.e. dynamical actions, both mechanical and electro-dynamic, can proceed forwards and backwards indifferently. This fundamental difference was first clearly recognized by R. Clausius, but it was a long time before it was generally appreciated, and even at the present time, experiments are continually being energetically conducted in the interests of the blending, to disprove the irreversibility.

They find their expression in the second law of thermodynamics, which states that in every thermo-chemical process the total entropy of the bodies concerned increases, and remains constant only in the ideal limiting case of reversible actions. The enormous importance of this theorem in heat and physical chemistry lay for some time in its peculiar contrast to the apparently insuperable difficulty of attacking it from a dynamical standpoint. It remained for L. Boltzmann to show a promising and, as it seemed, the only solution. Thereby were abandoned all claims to a purely dynamical explanation of the second law, and a purely statistical law was substituted, which embraced all the results of thermal and chemical measurements derived from an enormous number of isolated effects. While dynamical laws still hold good for the isolated quantities which represent the reactions between the atoms of a substance, so that for these the sign of the time variable is meaningless, the total magnitudes resulting from the inter-effects of a number of elementary actions form a basis for the laws of probability, which introduce into theoretical physics an element foreign to dynamics. This element is quite independent of dynamics, and, therefore, a new one.

From this point of view, the second law of thermo-dynamics appears solely as a law of probability, entropy as a measure of the probability, and the increase of entropy is equivalent to a statement that more probable events follow less probable ones. Thus the sign of the time variable is determined so that the later time is associated with the more probable event.

The characteristic of a theorem in probability is that it permits exceptions, and the determination of such exceptions forms an important problem in atomic statistical conception. The investigation of the conditions of equilibrium provides the most delicate proof of this. For while in dynamics equilibrium represents a condition of absolute invariability, statistical equilibrium is continually varying, more or less significant deviations referred to the so-called equilibrium of motion and, indeed, the magnitude of the mean deviation can be derived quantitatively and exactly from the theorems of probability. In this the statistical theory has established itself most brilliantly.

Most startling and most convincing to the unprejudiced investigator is the so-called Brownian movement, in which a fluid of uniform temperature and density at rest shows within it an incessant, extremely lively and confused motion of the small particles in suspension—a condition entirely inexplicable from the view-point of pure dynamics, but capable of attack, even to all the details of the calculation, by the statistical method.

Thus is being overcome the marked contrast between dynamics and thermo-dynamics, originally poles apart, principally by the acceptance of the assumption of absolute legality in all thermal and chemical observations, together with the introduction of the atomic methods which operate with a number of new natural constants, characteristic to themselves—the atomic weights. However, it appears that this is not the only, nor the most difficult, sacrifice which dynamics must offer if it wishes completely to embrace the theory of heat. It has probably not yet finished with the question of the discontinuity of matter. The laws of heat radiation, specific heat, electron emission, of radio-activity, and yet many other branches are in agreement that not only matter itself, but also the effects radiated from matter (if one can make such a distinction at all) possess discontinuous properties, which can again be characterized by a new natural constant: the elementary quantum of action. If that, too, is so infinitesimally small that the results of classical dynamics for all the greater effects will not be appreciably modified, then, taken as a foundation, it forms a part foreign to the constitution of the accepted theory. Its appearance at first was considered inconvenient, as not only does the significant meaning of the quantum of action almost preclude ease of representation, in contrast with electrons and atoms, which at any rate show certain analogies with celestial bodies, but also (and this is much more important) the place where the quantum of action should be introduced cannot be exactly defined. No wonder, therefore, that the classical theories even now oppose with all their might the recognition of this intruder, and that years must elapse before the dual assimilation process is complete.

There can be no doubt that the time will come when the chemical atomic weights, as well as the elementary quantum of action, whatever its name or form, will constitute an integral part of general dynamics. Then physical research will be unable to rest until the theory of heat and radiation has been welded into one united theory with mechanics and electro-dynamics.

The Nature of Light

One of the most important branches of work of this society (the *Kaiser-Wilhelm-Gesellschaft*) is the maintenance of a research laboratory for natural science. The society has, however, discovered the old truism that in its own sphere, as in all spheres of work, knowledge must precede application, and the more detailed our knowledge of any branch of physics, the richer and more lasting will be the results which we can draw from that knowledge.

In this respect, of all the branches of physics, there is no doubt that it is in optics that research work is most advanced, and, therefore, I am going to speak to you about the *Nature of Light*. I shall doubtless mention much that is familiar to each of you, but I shall also deal with newer problems still awaiting solution.

The first problem of physical optics, the condition necessary for the possibility of a true physical theory of light, is the analysis of all the complex phenomena connected with light, into objective and subjective parts. The first deals with those phenomena which are outside, and independent of, the organ of sight, the eye. It is the so-called light rays which constitute the domain of physical research. The second part embraces the inner phenomena, from eye to brain, and this leads us into the realms of physiology and psychology. It is not at all self-evident, from first principles, that the objective light rays can be completely separated from the sight sense, and that such a fundamental separation involves very difficult thinking cannot better be proved than by the following fact. Johann Wolfgang von Goethe was gifted with a very scientific mind (though little inclined to consider analytical methods), and would never see a detail without considering the whole, yet he definitely refused, a hundred years ago, to recognize this difference. Indeed, what assertion could give a greater impression of certainty to the unprejudiced than to say that light without the perceptive organ is inconceivable? But, the meaning of the word light in this connection, to give it an interpretation that is unassailable, is quite different from the light ray of the physicist. Though the name has been retained for simplicity, the physical theory of light or optics, in its most general sense, has as little to do with the eye and light perceptions as the theory of the pendulum has to do with sound perception. This ignoring of the sense-perceptions, this restricting to objective real phenomena, which doubtless, from the point of view of immediate interest, means a considerable sacrifice made to pure knowledge, has prepared a way for a great extension of the theory. This theory has surpassed all expectations, and yielded important results for the practical needs of mankind.

A very significant discovery relating to the physical nature of light rays was that light, emanating from stars or terrestrial sources, takes a certain measurable time to travel from the position of the source to the place at which it is observed. What is this something which spreads through empty space and moves through the atmosphere at the enormous speed of 300,000 kilometres per second? Isaac Newton, the founder of classical mechanics, made the most simple and obvious assumption that there are certain infinitesimally small corpuscles which are sent out in all directions with that velocity from a source of light, e.g. a glowing body. These particles are different for different colours. This provides a striking proof that a high authority can exercise a hindrance to the development of even this most exact of all natural sciences, for Newton's emanation theory was able to hold the field for a whole century, although another distinguished investigator, Christian Huygens, had from the first opposed it with his much more suitable undulation theory. Huygens did not place the velocity of light on a par with that of wind, as Newton did, but on a par with the velocity of sound, in which the velocity of propagation is something quite different from that of air movements. Consider the air surrounding a sounding instrument or the surface of water into which a stone has been thrown. It is not the air or water particles themselves that spread out in all directions with equal velocity, but the intensification and rarefaction, or wave crests and troughs; in other words, it is not with matter itself, but with a certain state of matter that we are concerned. To this end, Huygens formulated an ideal substance, uniformly occupying all space, as a foundation for his theory. This is the light-ether, the waves of which produce light perceptions in the eye, as air waves give rise to sound perceptions in the ear. The wave-length or frequency determines the colour in the same manner as it determines the pitch in sound. After a bitter controversy, Huygens's theory ultimately superseded that of Newton. This was due to the fact, amongst many others, that when two light rays of the same colour are superposed and made to travel on the same path, the intensities are not always simply additive, but under certain conditions the intensity is decreased and may even vanish. This last phenomenon, interference, can be straightway explained on Huygens's assumption that in every case the wave crests of one ray coincide with the wave troughs of the other ray. Newton's emanation theory naturally contradicts this, since it is impossible for two similar corpuscles travelling with the same speed in the same direction to neutralize one another.

A more significant fundamental view of the nature of light was obtained through the discovery of the identity of light and heat rays, and this was the first step on the way towards the complete separation of the science from the sense-perceptions. The cold light rays of the moon are physically of exactly the same nature as the black heat rays emitted from a stove, except that they are of much shorter wavelength. It is only natural that this assertion at first excited much discussion, and it is characteristic that Melloni, who played a great part in the verification of this fact, set out originally to disprove it. It must be remembered that here, as in all inductive results, a logical and conclusive proof cannot be given; it can only be shown that all laws which hold for light rays, namely those of reflection, refraction, interference, polarization, dispersion, emission, and absorption, are also true for heat rays. Whoever refuses to admit the identity of the two kinds of rays in spite of this, could certainly never be accused on this account of a logical fallacy; for he would always maintain that it is still possible in the future for an essential difference to be discovered. The practical weakness of his position is that he is, consequently, compelled to renounce a series of important conclusions immediately deduced from the theory of identity. He cannot, for example, maintain that moonbeams also carry heat, though this fact would, at present, appear indubitable to all rational physicists, though it has not been specifically proved.

Having accepted the identity of light and heat rays, there is no difficulty in connecting the infra-red rays with the chemically active ultra-violet rays at the other end of the spectrum. It was some time later that it was realized that this connection of different kinds of rays was capable of great extension, on both sides of the spectrum. Before such an advance could come about, as a preliminary, a transition from the mechanical to the electro-magnetic theory of light was necessary.

In spite of diversity of view, Newton, Huygens, and all their immediate successors were agreed that the clear understanding of the nature of light must be sought in the fundamentals of mechanical science, and this point of view was greatly stimulated by the strengthening of the mechanical theory of heat due to the discovery of the principle of conservation of energy. It is necessary for the explanation of polarization that ether oscillations are not longitudinal, moving in the direction of propagation, like air movements in a pipe, but are transversal, perpendicular to the direction of propagation, like those of a violin string. But one could get no nearer the nature of these oscillations from the laws of mechanics and elasticity. The more elaborate the hypotheses founded on the mechanical theory of light, whether ether was assumed to be continuous or atomic, the more evident became this inadequacy. At this stage, in the middle of the last century, came James Clerk Maxwell, with his bold hypothesis that light was electro-magnetic. His theory of electricity led him to the conclusion that every electrical disturbance moved from its source through space in waves with a velocity of 300,000 kilometres per second, and the coincidence of this figure, obtained from purely electrical measurements, with the magnitude of the velocity of light, led him to consider light as an electro-magnetic disturbance. The only proof of the correctness of this view lies in the fact that all deductions made from it agree with observation. The fundamental advance associated with his suggestion lies in the enormous simplification of the theory and in the number of results that can be immediately derived from it.

Now, the nature of electro-magnetic phenomena is no more intelligible than that of optical phenomena. To belittle the electro-magnetic theory of light, on the ground that it simply replaces one riddle by another, is to misunderstand the meaning of the theory. For its importance rests on the fact that it unites two branches of physics, which previously had to be treated as independent, and that, therefore, all theorems which are valid for one branch, are applicable to the other—a result which the mechanical theory of light did not, and could not, give. Before the introduction of the electro-magnetic theory, physics was divided into three separate branches—mechanics, optics, and electro-dynamics, and the unification of these is the ultimate and greatest aim of physical research. Though optics cannot be associated with mechanics, it combines completely with electro-dynamics, and thus the number of independent branches has been reduced to two—the penultimate step towards the unification of the physical world picture. When and how the last step will be made, the linking up of mechanics and electro-dynamics, cannot be said, and though many clever physicists are at present occupied with this question, the time does not yet seem ripe for the solution. However, the original mechanical comprehension of Nature, which will allow the coalescing of mechanics and electro-dynamics, has now been thrust into the background in the minds of most physicists, since it regards ether, or, if ether is not sufficient, some substitute as the medium of all electrical phenomena. That which has harmed it most is the result, deduced from Einstein's theory of relativity, that there can be no objective substantial ether, i.e. one independent of the observer. For, if that were not so, then when we consider two observers moving relative to one another in space, one at most could correctly assert that he was at rest relative to the ether, whereas, by the theory of relativity, each of the two could do so equally correctly.

What Maxwell could only prophecy, Heinrich Hertz was able to verify a generation later, when he showed how to produce the electro-magnetic waves calculated by Maxwell, and thereby ensured the final acceptance of the electro-magnetic theory of light, according to which electric waves only differ from heat and light rays in that they have very much greater wave-length. If the optical spectrum were extended on the side of the slow oscillations in a manner undreamt of at one time, the extension would be of equal importance with that made on the other side of the spectrum through the discovery of the Röntgen rays and the appreciably faster so-called Gamma rays of radio-active substances. These rays, too, have the character of light waves, and are electro-magnetic oscillations, but have a very much shorter wave-length. Laue's very recent discovery of interference phenomena with Röntgen rays has confirmed the belief that they obey the same laws. It is remarkable how simply and quietly the transition from the mechanical to the electro-magnetic theory was made in physical literature. This is a good example of the fact that the kernel of a physical theory is not the observations on which it is built, but the laws to which they give rise. The fundamental equations of optics remain unaltered: they have always been in agreement with observation, but they are no longer to be interpreted mechanically (although they were thus derived) but electro-magnetically, and this has increased enormously their range of application.

This is not the first time that an important goal has been reached by a path which has afterwards been proved to be untrustworthy. It would have been possible to seek a solution by supposing that the theory would have been better had it abstained, in general, from making special hypotheses, based on immediate observations, and to limit oneself to the pure facts, i.e. to the results of measurements. However, the theory would thus surrender the most important aid, absolutely necessary to its development, namely, the setting up and consistent expansion of ideas which lead to progress. For this, not only understanding, but also imagination is necessary. As it is, the mechanical theory of light has done its duty. Without it the present brilliant results of optics would not have been obtained so quickly.

Huygens's undulation theory has not been essentially altered by the electro-magnetic hypothesis, when it states that any disturbance spreads out from its source in concentric spherical waves. But it is electro-magnetic energy and not mechanical energy that is sent out, for an oscillating electric and magnetic field of force appears in place of periodic vibrations of the ether.

Considered from this advanced point of view, the study of light, or, as it is often more exactly called, the study of radiant energy, gives us a picture of a gigantic co-ordinated structure, unified and completed. In this, all electro-magnetic oscillations, though apparently of very different kinds, find their proper positions, and all are governed by the same laws of propagation, following Huygens's wave theory. On the one hand, we have the Hertzian waves a kilometre long; on the other, the hard Gamma rays, with many milliards of waves to the centimetre. The human eye has no place in this, it appears merely as an accidental and, although very delicate, a very limited piece of apparatus, for it can only perceive rays within a small spectral range of hardly an octave. Instead of the eye, special pieces of apparatus have been devised for receiving and measuring the different wave-lengths of the remainder of the spectrum. Such instruments are the wave detector, thermocouple, bolometer, radiometer, photographic plate, and the ionic cell. Thus, in optics, the separation of the physical foundations from the sense-perceptions has been accomplished in exactly the same way as in mechanics, where the conception of force has long lost its connection with the idea of muscular strength.

If I had delivered my lecture twenty years ago, I could have stopped here, for no further fundamental discoveries had then been made, and the imposing picture described above would have been a good conclusion which would have made modern physics famous. But probably I should not then have delivered this lecture, fearing that I should be able to present to you too little that was new. To-day it has become quite otherwise, for, since that time, the picture has been essentially changed. The proud structure, which I have just described to you, has recently revealed certain fundamental weaknesses, and not a few physicists maintain that new foundations are required already. The electro-magnetic theory must always remain untouched, but Huygens's wave theory is seriously threatened, at least in one essential detail, due to the discovery of certain new facts. Instead of collecting as many as possible of the multifarious facts available, I shall simple examine one of them in detail.

When ultra-violet rays fall on a piece of metal in a vacuum, a large number of electrons are shot off from the metal at a high velocity, and since the magnitude of this velocity does not essentially depend on the state of the metal, certainly not on its temperature, it is concluded that the energy of the electrons is not derived from the metal, but from the light rays which fall on the metal. This would not be strange in itself; it would even be assumed that the electro-magnetic energy of light waves is transformed into the kinetic energy of electronic movements. An apparently insuperable difficulty from the view of Huygens's wave theory is the fact (which was discovered by Philipp Lenard and others), that the velocity of the electrons does not depend on the intensity of the beam, but only on the wavelength, i.e. on the colour of light used. The velocity increases as the wave-length diminishes. If the distance between the metal and the source of light is continuously increased, using, for example, an electric spark as the source of light, the electrons continue to be flung off with the same velocity, in spite of the weakening of the illumination; the only difference is that the number of electrons thrown off per second decreases with the intensity of the light.

The difficulty is to state whence the electron obtains its energy, when the distance of the source of light becomes ultimately so great that the intensity of the light almost vanishes, and yet the electrons show no sign of diminution in their velocity. This must evidently be a case of a kind of accumulation of light energy at the spot from which the electron is flung out—an accumulation which is quite contrary to the uniform spreading out in all directions of electro-magnetic energy according to Huygens's wave theory. However, if it is assumed that the light source does not emit its rays uniformly but in impulses, something like an intermittent light, it follows that the energy of such a flash, spreading outwards in all directions in uniform waves, would finally be distributed over the surface of a sphere so large that the metal considered would receive but little of it. It is easy to calculate that under certain circumstances radiation extending for minutes, even hours, would be necessary for the liberation of one electron with the velocity corresponding to the colour of the light, while, in fact, no limiting condition can be determined, for the duration of radiation necessary to produce the effects; the action certainly takes place with great rapidity. Like ultra-violet rays, Röntgen rays and Gamma rays give us the same effect, though, owing to the very much shorter wave-lengths of these rays, the velocities of the liberated electrons are much greater.

The only possible explanation for these peculiar facts appears to be that the energy radiated from the source of light remains, not only for all time, but also throughout all space, concentrated in certain bundles, or, in other words, that light energy does not spread out quite uniformly in all directions, becoming continuously less intense, but always remains concentrated in certain definite quanta, depending only on the colour, and that these quanta move in all directions with the velocity of light. Such a light-quantum, striking the metal, communicates its energy to an electron, and the energy always remains the same, however great the distance from the source of light.

Here we have Newton's emanation theory resurrected in another and modified form. But interference, which was a bar to the further development of Newton's emanation theory, is also an enormous difficulty in the quantum theory of light, for it is difficult at present to see how two exactly similar light quanta, moving independently in space, and meeting on a common path, can neutralize each other, without violating the principle of energy.

From this state of affairs arose the pressing need of the radiation theory for an investigation to find some way out of this dilemma, difficult from all sides. A natural assumption to try is that the energy of the electrons driven off comes from the metal itself and not from the radiation, and, therefore, that the radiation acts merely as a liberator in the same manner that a small spark liberates any amount of energy in a powder cask. But the further assumption would be necessary that the amount of the energy freed depends solely on the manner in which it is freed. It is not difficult to point out somewhat analogous phenomena in other branches of physics. As an example, I will consider in greater detail a convenient illustration used by Max Born. Imagine a tall apple tree, all its branches weighed down with ripe fruit, all of the same size, but with stalks of different lengths; the apples are arranged so that those with short stalks are higher than those with long stalks. If an extremely weak, uniform wind blows through the branches, all the apples will oscillate slightly, without any one dropping, and the higher apples will oscillate more rapidly than the lower ones. If, now, the tree is shaken very gently with a definite rhythm, resonance will increase the oscillations of those apples whose period agrees with the period of the shaking, and a certain number of these will fall, the number increasing the longer and more forcibly the tree is shaken. These apples will reach the ground with a certain definite velocity determined only by their original height, i.e. by the lengths of their stalks; all the other apples remain on the tree.

It must be understood that this comparison, like every other, fails in many respects, since, in this illustration, the source of energy is not internal kinetic energy but gravitation. But the essential point is realized that the final velocity of the particles liberated depends solely on the period of the disturbance, while the intensity of the disturbance determines only the number of these particles.

Can one attribute, however, such a complicated structure and such a wealth of energy to a tiny piece of metal? This question is less awkward than would perhaps appear at first. For we have long known that the chemical atom is not by any means the simple invariable element of which all matter is constituted, but rather that every single atom, particularly one of a heavy metal, must be considered as a world in itself, and the farther one penetrates, the richer and more varied the structure appears. The energy contained in every gramme of a substance, according to the theory of relativity, amounts to over 20 billion calories, quite independently of its temperature—more than sufficient to liberate countless electrons.

Whether this presentation gives a possible way of saving the compromised wave theory, or simply leads ultimately to a blind alley, can only be settled by following the methods of research already outlined and seeing where they end. At this stage we must make use of theory. We must first of all examine more closely each of the two opposing hypotheses, without considering whether or not we have confidence in either of them, and must work out the results and reduce them to a form suitable for experimental verification. For this purpose, in addition to a training in physics and the requisite mathematical ability, it is necessary to have a discriminating judgment of the measure of the reliability that can be placed on the accuracy of the measurements; for the effects sought for are mostly of the same order as the errors of observation. It is not possible to-day to predict with certainty when any definite solution to this problem will be obtained.

What I have tried to set before you here about the action of light, holds in an exactly similar manner with regard to the cause of light, that is, to the phenomena of generation of light rays. In this also we have new riddles, difficult to unravel, which are at variance with certain surprisingly deep glimpses recently obtained into the laws governing natural phenomena. The only thing that can be said with certainty, is that the quanta, already referred to, play a characteristic part in connection with the origin of light.

According to the bold hypothesis of the Danish physicist Niels Bohr, the consequences of which have been astonishingly multiplied recently, electrons oscillate in every atom of an illuminated gas. These electrons circle about the nucleus in a greater or smaller number and at different distances, in certain definite paths and obey the same laws as those governing the motions of the planets about the sun. But light, arising from these oscillations, is not sent out from the atom into surrounding space uninterruptedly and uniformly, as are the sound waves from the prongs of a vibrating tuning-fork. The emission of light always takes place abruptly, by impulses, for it is not determined by the regular oscillations of the electrons themselves but is only emitted when these electron oscillations receive a sudden change and a certain disruption in themselves; a kind of internal catastrophe, which throws the electrons out of their original paths into others more stable but associated with less energy. It is the surplus amount of energy liberated by the atom which travels out into space as a light quantum.

Indeed, the most remarkable thing about this phenomenon is that the period of the emitted light, and therefore its colour, does not, in general, agree with the period of oscillation of the electrons, either in their original or in their final paths. It is definitely determined by the amount of energy emitted, since the more rapid the oscillations, the greater is the light quantum. It follows that a short wave-length corresponds to a large amount of energy, considered as a light quantum. If, therefore, for example, much energy is emitted, we get ultra-violet or even Röntgen rays; if, however, but little energy is emitted, red or infra-red rays result. It is at present a complete mystery why the oscillations of light produced in this way are, with the utmost regularity, strictly monochromatic.

Indeed, we might be inclined to consider all these ideas as the play of a vivid but empty imagination. When, on the other hand, we consider that these hypotheses help us to elucidate the mysterious structure of the spectra of the different chemical elements and, in particular, the complicated laws governing the spectral lines, not only as a whole but, as Arnold Sommerfeld first showed, partly even in minute details, with an exactness equal to, and even surpassing, that of the most accurate measurements—when we consider this we must, for good or ill, make up our minds to assign a real existence to these light quanta, at least at the instant of their origin.

What becomes of them later as light disperses—whether the energy of a quantum remains concentrated as in Newton's emanation theory or whether, as in Huygens's wave theory, it spreads out in all directions and gets less dense indefinitely—is another question of a very fundamental character, to which I have referred above.

So the present lecture on our knowledge of the physical nature of light ends, not in a proud proclamation, but in a modest question. In fact, this question, whether light rays themselves consist of quanta, or whether the quanta exist only in matter, is the chief and most difficult dilemma before which the whole quantum theory halts, and the answer to this question will be the first step towards further development.

The Origin and Development of the Quantum Theory

In this lecture I will endeavour to give a general account of the origin of the quantum theory, to sketch concisely its development up to the present, and to point out its immediate significance in physics.

Looking back over the last twenty years to the time when the conception and magnitude of the physical quantum of action first emerged from the mass of experimental facts, and looking back at the long and complicated path which finally led to an appreciation of its importance, the whole history of its development reminds me of the well-proved adage that "to err is human." And all the hard intellectual work of an industrious thinker must often appear vain and fruitless, but that striking occurrences sometimes provide him with an irrefutable proof of the fact that at the end of all his attempts, he does ultimately get one step nearer the truth. An indispensable hypothesis, though it does not guarantee a result, often arises from the pursuit of a definite object, the importance of which is not lessened by initial ill-success.

For me, such an object has, for a long time, been the solution of the problem of the distribution of energy in the normal spectrum of radiant heat. Gustav Kirchhoff showed that, in a space bounded by bodies at equal temperatures, but of arbitrary emissive and absorptive powers, the nature of the heat of radiation is completely independent of the nature of the bodies (1).* Later, a universal function was proved to exist, which depended only on temperature and wave-length, and was in no way related to the properties peculiar to any substance. The discovery of this remarkable function gave promise

* The numbers in parentheses refer to the notes at the end of the lecture.

of a deeper understanding of the relationship of energy to temperature, which forms the chief problem of thermo-dynamics, and, therefore, also of all molecular physics. There is no way at present available for obtaining this function but to select from all the various kinds of bodies occurring in Nature any one of known emission and absorption coefficients, and to calculate the heat radiation when the exchange of energy is stationary. According to Kirchhoff's theorem, this must be independent of the constitution of the body.

A body especially suited for this purpose appears to be Heinrich Hertz's oscillator, the laws of emission of which, for a given frequency, have recently been fully developed by Hertz (2). If a number of such oscillators be placed in a space enclosed by reflecting walls, they will exchange energy one with another by taking up or emitting electro-magnetic waves, analogous with a sound source and resonators, until finally stationary black radiation, so-called, obtains in the enclosure according to Kirchhoff's law. At one time I fostered the hope which seems to us rather naïve in these days, that the laws of classical electrodynamics, if applied sufficiently generally, and extended by suitable hypotheses, would be sufficient to explain the essential points of the phenomenon looked for, and to lead to the desired goal. To this end, I first of all developed the laws of emission and absorption of a linear resonator in the widest possible way, in fact, by a roundabout way which I could have avoided by using H. A. Lorentz's electron theory then complete in all fundamental points. But since I did not then fully believe in the electron hypothesis, I preferred to consider the energy flowing across a spherical surface of a certain radius enclosing the resonator. This only deals with phenomena in vacuo, but the knowledge of these is enough to enable us to draw the necessary conclusions about the energy changes of the resonator.

The result of this long series of investigations was the establishment of a general relation between the energy of a resonator of given period and the radiant energy of the corresponding region of the spectrum in the surrounding field when the energy exchange is stationary (3). Some of these investigations could be proved by comparison with available observations, particularly the damping measurements of V. Bjerknes, and this is a verification of the results (4). Thus the remarkable conclusion is reached that the relation does not depend on the nature of the resonator, in particular, not upon its damping coefficient—a very gratifying and welcome circumstance to me, since it allowed the whole problem to be simplified in so far that the energy of radiation could be replaced by the energy of the resonator. Thereby a system with one degree of freedom could be substituted for a complicated system with many degrees of freedom.

Indeed, this result was nothing but a step preparatory to starting on the real problem, which now appeared more formidable. The first attempt at solving the problem miscarried; for my original hope proved false, namely, that the radiation emitted from the resonator would, in some characteristic way, be distinct from the absorbed radiation and thus give a differential equation, by solving which it would be possible to derive a condition for the state of stationary radiation. The resonator only responded to the same rays as it emitted, and was not at all sensitive to neighbouring regions of the spectrum.

My assumption that the resonator could exert a one-sided, i.e. irreversible, effect on the energy of the surrounding field of radiation, was strongly contradicted by Ludwig Boltzmann (5). His mature experience led him to conclude that, according to the laws of classical mechanics, each phenomenon which I had considered, could operate in exactly the reverse direction. Thus, a spherical wave sent out from a resonator may be reversed and proceed in ever-diminishing concentric spheres until it shrinks up at the resonator and is absorbed by it, and causes again the energy previously absorbed to be emitted once more into space in the directions along which it had come. Even if, by introducing suitable limits, I could exclude from the hypothesis of "natural radiation" such singular phenomena as spherical waves travelling inwards, all these analyses show clearly that an essential connecting link is still missing for the complete understanding of the problem.

No other course remained open to me but to attack the problem from the opposite direction, namely, through thermo-dynamics, with which I felt more familiar. Here I was helped by my previous researches into the second law of thermo-dynamics, and I straightway conceived the idea of connecting the entropy and not the temperature of the resonator with the energy, indeed, not the entropy itself, but its second differential coefficient with respect to energy, since this has a direct physical meaning for the irreversibility of the exchange of energy between resonator and radiation. Since at that time I did not see my way clear to go any further into the dependence of entropy and probability, I could, first of all, only refer to results that had already been obtained. Now, in 1899, the most in-teresting result was the law of energy distribution which had just been discovered by W. Wien (6). The experimental proof of this was undertaken by F. Paschen at the *Hochschule*, Hanover, and by O. Lummer and E. Pringsheim at the *Reichsanstalt*, Charlottenburg. This law represents the dependence of the intensity of radiation on temperature by means of an exponen-tial function. Using this law to calculate the relation between the entropy and energy of a resonator, the remarkable result is obtained, that R, the reciprocal of the differential coefficient referred to above, is proportional to the energy (7). This exceedingly simple relation is a complete and adequate ex-pression of Wien's law of distribution of energy; for the depend-ence upon wave-length is always given immediately as well as the dependence upon energy by Wien's generally accepted law of displacements (8).

Since the whole problem deals with one of the universal laws of Nature, and since I believed then, as I do now, that the more general a natural law is, the simpler is its form (though it cannot always be said with certainty and finality which is the simpler form), I thought for a long time that the above relation, namely, that R is proportional to the energy, should be con-sidered as the foundation of the law of distribution of energy (9). This idea soon proved to be untenable in the light of more recent results. While Wien's law was confirmed for small values of energy, i.e. for short waves, O. Lummer and E. Pringsheim found large deviations in the case of long waves (10). Finally, the observations made by G. Rubens and F. Kurlbaum, with infra-red rays after transmission through fluorspar and rock salt (11), showed a totally different relation, which, under certain conditions, was still very simple. In this case, R is proportional, not to the energy, but to the square of the energy, and this relation is more accurate the larger the energies and wavelengths considered (12).

Thus, by direct experiment, two simple limits have been fixed for the function R, i.e. for small values of the energy it is proportional to the energy, for large values it is proportional to the square of the energy. It was obvious that in the general case the next step was to express R to the sum of two terms, one involving the first power, the other the second power of the energy, so that the first term was the predominating term for small values of the energy, the second term for large values. This gave a new formula for the radiation (13), which has stood the test of experiment fairly satisfactorily so far. No final exact experimental verification has yet been given and a new proof is badly needed (14).

If, however, the radiation formula should be shown to be absolutely exact, it would possess only a limited value, in the sense that it is a fortunate guess at an interpolation formula. Therefore, since it was first enunciated, I have been trying to give it a real physical meaning, and this problem led me to consider the relation between entropy and probability, along the lines of Boltzmann's ideas. After a few weeks of the most strenuous work of my life, the darkness lifted and an unexpected vista began to appear.

I will digress a little. According to Boltzmann, entropy is a measure of physical probability, and the essence of the second law of thermo-dynamics is that in Nature, the more often a condition occurs, the more probable it is. In Nature, entropy itself is never measured, but only the difference of entropy, and to this extent one cannot talk of absolute entropy without a certain arbitrariness. Yet, the introduction of an absolute magnitude of entropy, suitably defined, is allowed, since certain general theorems can be expressed very simply by doing so. As far as I can see, it is exactly the same with energy. Energy itself cannot be measured, but only a difference of energy. Therefore, one did not previously deal with energy, but with work, and Ernst Mach, who was concerned to a great extent with the conservation of energy, but avoided all speculations outside the domain of observation, has always refrained from talking of energy itself. Similarly, at first in thermo-chemistry, one considered heat of reaction, i.e. difference of energy, until William Ostwald emphatically showed that many involved considerations could be very much simplified, if one dealt with energy itself instead of calorimetric values. The undetermined additive constant in the expression for energy was fixed later by the relativity theorem of the relation between energy and inertia (15).

As in the case of energy, we can define absolute value for entropy and consequently for physical probability, if the additive constant is fixed so that entropy and energy vanish simultaneously. (It would be better to substitute temperature for energy here.) On this basis a comparatively simple combinatory method was derived for calculating the physical probability of a certain distribution of energy in a system of resonators. This method leads to the same expression for entropy as was obtained from the radiation theory (16). As an offset against much disappointment, I derived much satisfaction from the fact that Ludwig Boltzmann, in a letter acknowledging my paper, gave me to understand that he was interested in, and fundamentally in agreement with, my ideas.

For numerical applications of this method of probability we require two universal constants, each of which has an independent physical significance. The supplementary calculation of these constants from the radiation theory shows whether the method is merely a numerical one or has an actual physical meaning. The first constant is of a more or less formal nature, it depends on the definition of temperature. The value of this constant is $\frac{2}{3}$ if temperature be defined as the mean kinetic energy of a molecule in an ideal gas, and is, therefore, a very small quantity (17). With the conventional measure of temperature, however, this constant has an extremely small value, which is naturally closely dependent upon the energy of a single molecule, and an exact knowledge of it leads, therefore, to the calculation of the mass of a molecule and the quantities depending upon it. This constant is frequently called Boltzmann's constant, though Boltzmann himself, to my knowledge, never introduced it—a curious circumstance, explained by the fact that Boltzmann, as appears from various remarks by him (18), never thought of the practicability of measuring this constant exactly. Nothing can better illustrate the impetuous advance made in experimental methods in the last twenty years than the fact that since then, not one only, but a whole series of methods have been devised for measuring the mass of a single molecule with almost the same accuracy as that of a planet.

While, at the time that I carried out the corresponding calculations from the radiation theory, it was impossible to verify exactly the figure obtained, and all that could be achieved was to check the order of magnitude; shortly afterwards, E. Rutherford and H. Geiger (19), succeeded in determining the value of the elementary electric charge to be $4 \cdot 65 \times 10^{-10}$ electro-static units, by directly counting a-particles. The agreement of this figure with that calculated by me, $4 \cdot 69 \times 10^{-10}$, was a definite confirmation of the usefulness of my theory. Since then, more perfect methods have been developed by E. Regener, R. A. Millikan, and others (20), and have given a value slightly higher than this.

The interpretation of the second universal constant of the radiation formula was much less simple. I called it the elementary quantum of action, since it is a product of energy and time, and was calculated to be $6 \cdot 55 \times 10^{-27}$ erg. sec. Though it was indispensable for obtaining the right expression for entropy—for it is only by the help of it that the magnitude of the standard element of probability could be fixed for the probability calculations (21)—it proved itself unwieldy and cumbrous in all attempts to make it fit in with classical theory in any form. So long as this constant could be considered infinitesimal, as when dealing with large energies or long periods of time, everything was in perfect agreement, but in the general case, a rift appeared, which became more and more pronounced the weaker and more rapid the oscillations considered. The failure of all attempts to bridge this gap soon showed that undoubtedly one of two alternatives must obtain. Either the quantum of action was a fictitious quantity, in which case all the deductions from the radiation theory were largely illusory and were nothing more than mathematical juggling. Or the radiation theory is founded on actual physical ideas, and then the quantum of action must play a fundamental rôle in physics, and proclaims itself as something quite new and hitherto unheard of, forcing us to recast our physical ideas, which, since the foundation of the infinitesimal calculus by Leibniz and Newton, were built on the assumption of continuity of all causal relations.

Experience has decided for the second alternative. That this decision should be made so soon and so certainly is not due to the verification of the law of distribution of energy in heat radiation, much less to my special derivation of this law, but to the restless, ever-advancing labour of those workers who have made use of the quantum of action in their investigations.

The first advance in this work was made by A. Einstein, who proved, on the one hand, that the introduction of the energy quanta, required by the quantum of action, appeared suitable for deriving a simple explanation for a series of remarkable observations of light effects, such as Stokes's rule, emission of electrons, and ionization of gases (22). On the other hand, by identifying the energy of a system of resonators with the energy of a rigid body, he derived a formula for the specific heat of a rigid body, which gives again quite correctly the variation of specific heat, particularly its decrease with decrease of temperature (23). It is not my duty here to give even an approximately complete account of this work. I can only point out the most important characteristic stages in the progress of knowledge.

We will now consider problems in heat and chemistry. As far as the specific heat of a solid body is concerned, Einstein's method, based on the assumption of a single characteristic oscillation of the atom, has been extended by M. Born and Th. von Kármán to the case of various characteristic oscillations, more in agreement with practice (24). By greatly simplifying the assumptions regarding the nature of the oscillations, P. Debye obtained a comparatively simple formula for the specific heat of a solid body (25). This not only corroborates, particularly for low temperatures, the experimental values obtained by W. Nernst and his school, but also is in good agreement with the elastic and optical properties of the body. Further, quantum effects are very noticeable when considering the specific heat of gases. W. Nernst had shown at an early stage (26) that the quantum of energy of an oscillation must correspond to the quantum of energy of a rotation, and accordingly expected that the energy of rotation of a gas molecule would decrease with temperature. A. Eucken's measurements of the specific heat of hydrogen verified this deduction (27), and the fact that the calculations of A. Einstein and O. Stern, P. Ehrenfest, and others have not yet been in satisfactory agreement can be ascribed to our incomplete knowledge of the form of the hydrogen molecule. The work of N. Bjerrum, E. v. Bahr, H. Rubens, and G. Hettner, etc., on absorption bands in the infrared rays, shows that there can be no doubt that the rotations of the gas molecules indicated by the quantum conditions do actually exist. However, no one has yet succeeded in giving a complete explanation of these remarkable rotations.

Since all the affinity of a substance is ultimately bound up with its entropy, the theoretical calculation of entropy by means of quanta gives a method of attacking all problems in chemical affinity. Nernst's chemical constant is a characteristic for the absolute value of the entropy of a gas. O. Sackur calculated this constant directly (28) by a combinatory method similar to my method with oscillators, while O. Stern and H. Tetrode, by careful examination of experimental data of evaporation, determined the difference of the entropies of gaseous and non-gaseous substances (29).

The cases considered so far deal with thermo-dynamical equilibrium, which only give statistical mean values for a number of particles and long periods of time. This observation of electronic impulses, however, leads directly to the dynamical details of the phenomena considered. The determination by J. Franck and G. Hertz of the so-called resonance potential, or that critical velocity, the minimum velocity which an electron must have to bring about the emission of a quantum of light by collision with a neutral atom, is as direct a method of measuring the quantum of action as can be desired (30). Also, in the case of the characteristic radiation of the Röntgen spectrum discovered by C. G. Barkla, similar methods which gave very good results were developed by D. L. Webster, E. Wagner, and others.

The liberation of quanta of light by electronic impulses is the converse of the emission of electrons by projection of light, Röntgen or Gamma rays, and here, again, the quanta of energy determined from the quantum of action and the frequency of oscillations play a characteristic part in the same way as we have seen above, in that the velocity of the electrons emitted does not depend on the intensity of the radiation (31), but on the wavelength of the light emitted (32). From a quantitative point of view, also, Einstein's relations for light quanta mentioned above have been verified in every way, particularly by R. A. Millikan, who determined the initial velocities of the emitted electrons (33), while the significance of the light quantum in causing photo-chemical reactions has been made clear by E. Warburg (34).

The results quoted above, collected from the most varied branches of physics, present an overwhelming case for the existence of the quantum of action, and the quantum hypothesis was put on a very firm foundation by Niels Bohr's theory of the atom. This theory was destined, by means of the quantum of action, to open a door into the wonderland of spectroscopy, which had obstinately defied all investigators since the discovery of spectral analysis. Once the way was made clear, a mass of new knowledge was obtained concerning this branch of science, as well as allied branches of physics and chemistry. The first brilliant result was Balmer's series for hydrogen and helium, including the reduction of the universal Rydberg constants to pure numbers (35), by which the small difference between hydrogen and helium was found to be due to the slower motion of the heavier atomic core. This led immediately to the investigation of other series in the optical and Röntgen spectra by means of Ritz's useful combination principle, the fundamental meaning of which was now demonstrated for the first time.

In the face of these numerous verifications (which could be

considered as very strong proofs in view of the great accuracy of spectroscopic measurements), those who had looked on the problem as a game of chance were finally compelled to throw away all doubt when A. Sommerfeld showed that—by extending the laws of distribution of quanta to systems with several degrees of freedom (and bearing in mind the variability of mass according to the theory of relativity)—an elegant formula follows which must, so far as can be determined by the most delicate measurements now possible (those of F. Paschen (36)), solve the riddle of the structure of hydrogen and helium spectra (37). This is an accomplishment in every way comparable with the famous discovery of the planet Neptune, whose existence and position had been calculated by Leverrier before it had been seen by human eye. Proceeding further along the same lines, P. Epstein succeeded in giving a complete explanation of the Stark effect of the electrical separation of the spectral lines (38), and P. Debye in giving a simple meaning to the K-series of the Röntgen spectrum, investigated by Manne Siegbahn (39). Moreover, there followed a large number of wider investigations, which explained more or less successfully the mystery of the structure of the atom.

In view of all these results—a complete explanation would involve the inclusion of many more well-known names—an unbiased critic must recognize that the quantum of action is a universal physical constant, the value of which has been found from several very different phenomena to be 6.54×10^{-27} ergs. secs. (40). It must seem a curious coincidence that at the time when the idea of general relativity is making headway and leading to unexpected results, Nature has revealed, at a point where it could be least foreseen, an absolute invariable unit, by means of which the magnitude of the action in a time space element can be represented by a definite number, devoid of ambiguity, thus eliminating the hitherto relative character.

Yet no actual quantum theory has been formed by the introduction of the quantum of action. But perhaps this theory is not so far distant as the introduction of Maxwell's light theory was from the discovery of the velocity of light by Olaf Römer. The difficulties in the way of introducing the quantum of action into classical theory from the beginning have been mentioned above. As years have elapsed, these difficulties have increased rather than diminished, and although the impetuous advance of research has dealt with some of them, yet the inevitable gaps remaining in any extension are all the more painful to the conscientious and systematic worker. That which serves as the foundation of the law of action in Bohr's theory is made up of certain hypotheses which were flatly rejected, without any question, a generation ago by physicists. That quite definite orbits determined by quanta are a special feature of the atom may be considered admissible, but it is less easy to assume that the electrons, moving in these paths with a definite acceleration, radiate no energy. But that the quite sharply defined frequency of an emitted light quantum should be different from the frequency of the emitted electrons must seem, at first sight, to a physicist educated in the classical school, an almost unreasonable demand on his imagination.

However, figures are decisive, and the conclusion is that things have been gradually reversed. At first a new foreign element was fitted into a structure, generally considered fixed, with as little change as possible; but now the intruder, after gaining a secure place for itself, has taken the offensive, and to-day it is almost certain that it will undermine the old structure in some way or other. The question is at what place and to what degree this will happen.

If a surmise be allowed as to the probable outcome of this struggle, everything seems to indicate that the great principles of thermo-dynamics, derived from the classical theory, will not only maintain their central position in the quantum theory, but will be greatly extended. The adiabatic hypothesis of P. Ehrenfest (41) plays the same part in the quantum theory as the original experiments played in the founding of classical thermo-dynamics. Just as R. Clausius introduced, as a basis for the measure of entropy, the theorem that any two conditions of a material system are transformable one to the other by reversible processes, so Bohr's new ideas showed the corresponding way to explore the problems opened up by him.

A question, from the complete answer to which we may

expect far-reaching explanations, is what becomes of the energy of a light quantum after perfect emission? Does it spread out, as it progresses, in all directions, as in Huygens's wave theory, and while covering an ever-larger amount of space, diminish without limit? Or does it travel along as in Newton's emanation theory like a projectile in one direction? In the first case the quantum could never concentrate its energy in a particular spot to enable it to liberate an electron from the atomic influences; in the second case we would have the complete triumph of Maxwell's theory, and the continuity between static and dynamic fields must be sacrificed, and with it the present complete explanation of interference phenomena, which have been investigated in all details. Both these alternatives would have very unpleasant consequences for the modern physicist.

In each case there can be no doubt that science will be able to overcome this serious dilemma, and that what seems now to be incompatible may later be regarded as most suitable on account of its harmony and simplicity. Until this goal is attained the problem of the quantum of action will not cease to stimulate research and to yield results, and the greater the difficulties opposed to its solution, the greater will be its significance for the extension and deepening of all our knowledge of physics.

Notes

*(The Bibliography is by no means complete, but serves
as an indication of the papers which bear on the subject.)*

1. G. Kirchhoff. Über das Verhältnis zwischen dem Emissions-
   vermögen und dem Absorptionsvermögen der Körper
   für Wärme und Licht. Gesammelte Abhandlungen.
   Leipzig, J. A. Barth, 1882, p. 597 (§ 17).
2. H. Hertz. Ann. d. Phys., **36**, p. 1, 1889.
3. Sitz.-Ber. d. Preuss. Akad. d. Wiss., 18 May, 1899, p. 455.
4. Sitz.-Ber. d. Preuss. Akad. d. Wiss., 20 Feb., 1896. Also
   Ann. d. Phys., **60**, p. 577. 1897.
5. L. Boltzmann. Sitz.-Ber. d. Preuss. Akad. d. Wiss., 3 March,
   1898, p. 182.
6. W. Wien. Ann. d. Phys., **58**, p. 662, 1896.
7. According to Wien's Law of Energy Distribution, the relation between the energy U of a resonator and the temperature is given by

$$U = ae^{-\frac{b}{T}}.$$

If S denote the entropy, then

$$\frac{1}{T} = \frac{dS}{dU}$$

and R of the text $= 1 \Big/ \dfrac{d^2S}{dU^2} = -bU.$

8. Wien's law of displacements shows that the energy U of a
   resonator $= \nu f\left(\dfrac{T}{\nu}\right)$ where ν is the frequency of the
   oscillations.
9. Ann. d. Phys., **1**, p. 719, 1900.

10. O. Lummer and E. Pringsheim. Verh. d. Deutsch. Phys. Ges., **2,** p. 163, 1900.
11. H. Rubens and F. Kurlbaum. Sitz.-Ber. der Preuss. Akad. d. Wiss., 25 October, 1900, p. 929.
12. For large values of T, the investigations of H. Rubens and F. Kurlbaum give U = cT. Then from (7)

$$R = 1 \Big/ \frac{d^2S}{\nu U^2} = - \frac{U^2}{c}.$$

13. If R be assumed to be equal to

$$1 \Big/ \frac{d^2S}{dU^2} = - bU - \frac{U^2}{c}$$

then it follows by integration that

$$\frac{1}{T} = \frac{dS}{dU} = \frac{1}{b} \cdot \log \left(1 + \frac{bc}{U}\right)$$

and hence the radiation formula

$$U = \frac{bc}{e^{\frac{b}{T}} - 1}.$$

Cf. Verh. d. Deutsch. Phys. Ges., 19 October, 1900, p. 202.

14. Cf. W. Nernst and Th. Wulf. Verh. d. Deutsch. Phys. Ges., **21,** p. 294, 1919.
15. The absolute value of the energy is equal to the product of the mass and the square of the velocity of light.
16. Verh. d. Deutsch. Phys. Ges., 14 December, 1900, p. 237.
17. In general, if k be the first radiation constant, the mean kinetic energy of a gas molecule

$$U = \tfrac{2}{3}k T.$$

If T = U, $k = \tfrac{2}{3}$. In the case of Kelvin's absolute temperature scale, T is defined by putting the difference of temperature at the boiling and freezing-points of water equal to 100.

18. Cf. e.g. L. Boltzmann. Zur Erinnerung an Josef Loschmidt. Populäre Schriften, p. 245, 1905.

19. E. Rutherford and H. Geiger. Proc. Roy. Soc., A., Vol. 81, p. 162, 1908.

20. Cf. R. A. Millikan. Phys. Zeitschr., **14,** p. 796, 1913.

21. The calculation of the probability of a physical state consists of an enumeration of a finite number of equally probable individual cases through which such a state is realized. In order to differentiate these individual cases from one another, it is necessary to fix definitely the nature of each individual case.

22. A. Einstein. Ann. d. Phys., **17,** p. 132, 1905.

23. A. Einstein. Ann. d. Phys., **22,** p. 180, 1907.

24. M. Born and Th. v. Kármán. Phys. Zeitschr., **14,** p. 15, 1913.

25. P. Debye. Ann. d. Phys., **39,** p. 789, 1912.

26. W. Nernst. Phys. Zeitschr., **13,** p. 1064, 1912.

27. A. Euchen. Sitz.-Ber. d. Preuss. Akad. d. Wiss., p. 141, 1912.

28. O. Sackur. Ann. d. Phys., **36,** p. 958, 1911.

29. O. Stern. Phys. Zeitschr., **14,** p. 629, 1913.
    H. Tetrode. Ber. d. Akad. d. Wiss. v. Amsterdam, 27 February and 27 March, 1915.

30. J. Franck and G. Hertz. Verh. d. Deutsch. Phys. Ges., **16,** p. 512, 1914.

31. Ph. Lenard. Ann. d. Phys., **8,** p. 149, 1902.

32. E. Ladenburg. Verh. d. Deutsch. Phys. Ges., **9,** p. 504, 1907.

33. R. A. Millikan. Phys. Zeitschr., **17,** p. 217, 1916.

34. E. Warburg. Über den Energieumsatz bei photochemischen Vorgängen in Gasen. Sitz.-Ber. d. Preuss. Akad. d. Wiss., from 1911 onwards.

35. N. Bohr. Phil. Mag., **30,** p. 394, 1915.

36. F. Paschen. Ann. d. Phys., **50,** p. 901, 1916.

37. A. Sommerfeld. Ann. d. Phys., **51,** pp. 1, 125, 1916.

38. P. Epstein. Ann. d. Phys., **50,** p. 489, 1916.

39. P. Debye. Phys. Zeitschr., **18,** p. 276, 1917.

40. E. Wagner. Ann. d. Phys., **57,** p. 467, 1918.
    R. Ladenburg. Jahr. d. Radioaktivität u. Elektronik, **17,** p. 144, 1920.

41. P. Ehrenfest. Ann. d. Phys., **51,** p. 327, 1916.

A CATALOG OF SELECTED

DOVER BOOKS
IN SCIENCE AND MATHEMATICS

A CATALOG OF SELECTED
DOVER BOOKS
IN SCIENCE AND MATHEMATICS

QUALITATIVE THEORY OF DIFFERENTIAL EQUATIONS, V.V. Nemytskii and V.V. Stepanov. Classic graduate-level text by two prominent Soviet mathematicians covers classical differential equations as well as topological dynamics and ergodic theory. Bibliographies. 523pp. 5⅜ × 8½. 65954-2 Pa. $10.95

MATRICES AND LINEAR ALGEBRA, Hans Schneider and George Phillip Barker. Basic textbook covers theory of matrices and its applications to systems of linear equations and related topics such as determinants, eigenvalues and differential equations. Numerous exercises. 432pp. 5⅜ × 8½. 66014-1 Pa. $9.95

QUANTUM THEORY, David Bohm. This advanced undergraduate-level text presents the quantum theory in terms of qualitative and imaginative concepts, followed by specific applications worked out in mathematical detail. Preface. Index. 655pp. 5⅜ × 8½. 65969-0 Pa. $13.95

ATOMIC PHYSICS (8th edition), Max Born. Nobel laureate's lucid treatment of kinetic theory of gases, elementary particles, nuclear atom, wave-corpuscles, atomic structure and spectral lines, much more. Over 40 appendices, bibliography. 495pp. 5⅜ × 8½. 65984-4 Pa. $12.95

ELECTRONIC STRUCTURE AND THE PROPERTIES OF SOLIDS: The Physics of the Chemical Bond, Walter A. Harrison. Innovative text offers basic understanding of the electronic structure of covalent and ionic solids, simple metals, transition metals and their compounds. Problems. 1980 edition. 582pp. 6⅛ × 9¼. 66021-4 Pa. $15.95

BOUNDARY VALUE PROBLEMS OF HEAT CONDUCTION, M. Necati Özisik. Systematic, comprehensive treatment of modern mathematical methods of solving problems in heat conduction and diffusion. Numerous examples and problems. Selected references. Appendices. 505pp. 5⅜ × 8½. 65990-9 Pa. $11.95

A SHORT HISTORY OF CHEMISTRY (3rd edition), J.R. Partington. Classic exposition explores origins of chemistry, alchemy, early medical chemistry, nature of atmosphere, theory of valency, laws and structure of atomic theory, much more. 428pp. 5⅜ × 8½. (Available in U.S. only) 65977-1 Pa. $10.95

A HISTORY OF ASTRONOMY, A. Pannekoek. Well-balanced, carefully reasoned study covers such topics as Ptolemaic theory, work of Copernicus, Kepler, Newton, Eddington's work on stars, much more. Illustrated. References. 521pp. 5⅜ × 8½. 65994-1 Pa. $12.95

PRINCIPLES OF METEOROLOGICAL ANALYSIS, Walter J. Saucier. Highly respected, abundantly illustrated classic reviews atmospheric variables, hydrostatics, static stability, various analyses (scalar, cross-section, isobaric, isentropic, more). For intermediate meteorology students. 454pp. 6⅛ × 9¼. 65979-8 Pa. $14.95

RELATIVITY, THERMODYNAMICS AND COSMOLOGY, Richard C. Tolman. Landmark study extends thermodynamics to special, general relativity; also applications of relativistic mechanics, thermodynamics to cosmological models. 501pp. 5⅜ × 8½. 65383-8 Pa. $12.95

APPLIED ANALYSIS, Cornelius Lanczos. Classic work on analysis and design of finite processes for approximating solution of analytical problems. Algebraic equations, matrices, harmonic analysis, quadrature methods, much more. 559pp. 5⅜ × 8½. 65656-X Pa. $12.95

SPECIAL RELATIVITY FOR PHYSICISTS, G. Stephenson and C.W. Kilmister. Concise elegant account for nonspecialists. Lorentz transformation, optical and dynamical applications, more. Bibliography. 108pp. 5⅜ × 8½. 65519-9 Pa. $4.95

INTRODUCTION TO ANALYSIS, Maxwell Rosenlicht. Unusually clear, accessible coverage of set theory, real number system, metric spaces, continuous functions, Riemann integration, multiple integrals, more. Wide range of problems. Undergraduate level. Bibliography. 254pp. 5⅜ × 8½. 65038-3 Pa. $7.95

INTRODUCTION TO QUANTUM MECHANICS With Applications to Chemistry, Linus Pauling & E. Bright Wilson, Jr. Classic undergraduate text by Nobel Prize winner applies quantum mechanics to chemical and physical problems. Numerous tables and figures enhance the text. Chapter bibliographies. Appendices. Index. 468pp. 5⅜ × 8½. 64871-0 Pa. $11.95

ASYMPTOTIC EXPANSIONS OF INTEGRALS, Norman Bleistein & Richard A. Handelsman. Best introduction to important field with applications in a variety of scientific disciplines. New preface. Problems. Diagrams. Tables. Bibliography. Index. 448pp. 5⅜ × 8½. 65082-0 Pa. $12.95

MATHEMATICS APPLIED TO CONTINUUM MECHANICS, Lee A. Segel. Analyzes models of fluid flow and solid deformation. For upper-level math, science and engineering students. 608pp. 5⅜ × 8½. 65369-2 Pa. $13.95

ELEMENTS OF REAL ANALYSIS, David A. Sprecher. Classic text covers fundamental concepts, real number system, point sets, functions of a real variable, Fourier series, much more. Over 500 exercises. 352pp. 5⅜ × 8½. 65385-4 Pa. $10.95

PHYSICAL PRINCIPLES OF THE QUANTUM THEORY, Werner Heisenberg. Nobel Laureate discusses quantum theory, uncertainty, wave mechanics, work of Dirac, Schroedinger, Compton, Wilson, Einstein, etc. 184pp. 5⅜ × 8½. 60113-7 Pa. $5.95

INTRODUCTORY REAL ANALYSIS, A.N. Kolmogorov, S.V. Fomin. Translated by Richard A. Silverman. Self-contained, evenly paced introduction to real and functional analysis. Some 350 problems. 403pp. 5⅜ × 8½. 61226-0 Pa. $9.95

PROBLEMS AND SOLUTIONS IN QUANTUM CHEMISTRY AND PHYSICS, Charles S. Johnson, Jr. and Lee G. Pedersen. Unusually varied problems, detailed solutions in coverage of quantum mechanics, wave mechanics, angular momentum, molecular spectroscopy, scattering theory, more. 280 problems plus 139 supplementary exercises. 430pp. 6½ × 9¼. 65236-X Pa. $12.95

ASYMPTOTIC METHODS IN ANALYSIS, N.G. de Bruijn. An inexpensive, comprehensive guide to asymptotic methods—the pioneering work that teaches by explaining worked examples in detail. Index. 224pp. 5⅜ × 8½.   64221-6 Pa. $6.95

OPTICAL RESONANCE AND TWO-LEVEL ATOMS, L. Allen and J.H. Eberly. Clear, comprehensive introduction to basic principles behind all quantum optical resonance phenomena. 53 illustrations. Preface. Index. 256pp. 5⅜ × 8½.
65533-4 Pa. $7.95

COMPLEX VARIABLES, Francis J. Flanigan. Unusual approach, delaying complex algebra till harmonic functions have been analyzed from real variable viewpoint. Includes problems with answers. 364pp. 5⅜ × 8½.   61388-7 Pa. $8.95

ATOMIC SPECTRA AND ATOMIC STRUCTURE, Gerhard Herzberg. One of best introductions; especially for specialist in other fields. Treatment is physical rather than mathematical. 80 illustrations. 257pp. 5⅜ × 8½.   60115-3 Pa. $5.95

APPLIED COMPLEX VARIABLES, John W. Dettman. Step-by-step coverage of fundamentals of analytic function theory—plus lucid exposition of five important applications: Potential Theory; Ordinary Differential Equations; Fourier Transforms; Laplace Transforms; Asymptotic Expansions. 66 figures. Exercises at chapter ends. 512pp. 5⅜ × 8½.   64670-X Pa. $11.95

ULTRASONIC ABSORPTION: An Introduction to the Theory of Sound Absorption and Dispersion in Gases, Liquids and Solids, A.B. Bhatia. Standard reference in the field provides a clear, systematically organized introductory review of fundamental concepts for advanced graduate students, research workers. Numerous diagrams. Bibliography. 440pp. 5⅜ × 8½.   64917-2 Pa. $11.95

UNBOUNDED LINEAR OPERATORS: Theory and Applications, Seymour Goldberg. Classic presents systematic treatment of the theory of unbounded linear operators in normed linear spaces with applications to differential equations. Bibliography. 199pp. 5⅜ × 8½.   64830-3 Pa. $7.95

LIGHT SCATTERING BY SMALL PARTICLES, H.C. van de Hulst. Comprehensive treatment including full range of useful approximation methods for researchers in chemistry, meteorology and astronomy. 44 illustrations. 470pp. 5⅜ × 8½.   64228-3 Pa. $10.95

CONFORMAL MAPPING ON RIEMANN SURFACES, Harvey Cohn. Lucid, insightful book presents ideal coverage of subject. 334 exercises make book perfect for self-study. 55 figures. 352pp. 5⅜ × 8¼.   64025-6 Pa. $9.95

OPTICKS, Sir Isaac Newton. Newton's own experiments with spectroscopy, colors, lenses, reflection, refraction, etc., in language the layman can follow. Foreword by Albert Einstein. 532pp. 5⅜ × 8½.   60205-2 Pa. $9.95

GENERALIZED INTEGRAL TRANSFORMATIONS, A.H. Zemanian. Graduate-level study of recent generalizations of the Laplace, Mellin, Hankel, K. Weierstrass, convolution and other simple transformations. Bibliography. 320pp. 5⅜ × 8½.   65375-7 Pa. $8.95

THE ELECTROMAGNETIC FIELD, Albert Shadowitz. Comprehensive undergraduate text covers basics of electric and magnetic fields, builds up to electromagnetic theory. Also related topics, including relativity. Over 900 problems. 768pp. 5⅜ × 8¼. 65660-8 Pa. $18.95

FOURIER SERIES, Georgi P. Tolstov. Translated by Richard A. Silverman. A valuable addition to the literature on the subject, moving clearly from subject to subject and theorem to theorem. 107 problems, answers. 336pp. 5⅜ × 8½. 63317-9 Pa. $8.95

THEORY OF ELECTROMAGNETIC WAVE PROPAGATION, Charles Herach Papas. Graduate-level study discusses the Maxwell field equations, radiation from wire antennas, the Doppler effect and more. xiii + 244pp. 5⅜ × 8½. 65678-0 Pa. $6.95

DISTRIBUTION THEORY AND TRANSFORM ANALYSIS: An Introduction to Generalized Functions, with Applications, A.H. Zemanian. Provides basics of distribution theory, describes generalized Fourier and Laplace transformations. Numerous problems. 384pp. 5⅜ × 8½. 65479-6 Pa. $9.95

THE PHYSICS OF WAVES, William C. Elmore and Mark A. Heald. Unique overview of classical wave theory. Acoustics, optics, electromagnetic radiation, more. Ideal as classroom text or for self-study. Problems. 477pp. 5⅜ × 8½. 64926-1 Pa. $12.95

CALCULUS OF VARIATIONS WITH APPLICATIONS, George M. Ewing. Applications-oriented introduction to variational theory develops insight and promotes understanding of specialized books, research papers. Suitable for advanced undergraduate/graduate students as primary, supplementary text. 352pp. 5⅜ × 8½. 64856-7 Pa. $8.95

A TREATISE ON ELECTRICITY AND MAGNETISM, James Clerk Maxwell. Important foundation work of modern physics. Brings to final form Maxwell's theory of electromagnetism and rigorously derives his general equations of field theory. 1,084pp. 5⅜ × 8½. 60636-8, 60637-6 Pa., Two-vol. set $19.90

AN INTRODUCTION TO THE CALCULUS OF VARIATIONS, Charles Fox. Graduate-level text covers variations of an integral, isoperimetrical problems, least action, special relativity, approximations, more. References. 279pp. 5⅜ × 8½. 65499-0 Pa. $7.95

HYDRODYNAMIC AND HYDROMAGNETIC STABILITY, S. Chandrasekhar. Lucid examination of the Rayleigh-Benard problem; clear coverage of the theory of instabilities causing convection. 704pp. 5⅜ × 8¼. 64071-X Pa. $14.95

CALCULUS OF VARIATIONS, Robert Weinstock. Basic introduction covering isoperimetric problems, theory of elasticity, quantum mechanics, electrostatics, etc. Exercises throughout. 326pp. 5⅜ × 8½. 63069-2 Pa. $7.95

DYNAMICS OF FLUIDS IN POROUS MEDIA, Jacob Bear. For advanced students of ground water hydrology, soil mechanics and physics, drainage and irrigation engineering and more. 335 illustrations. Exercises, with answers. 784pp. 6⅛ × 9¼. 65675-6 Pa. $19.95

NUMERICAL METHODS FOR SCIENTISTS AND ENGINEERS, Richard Hamming. Classic text stresses frequency approach in coverage of algorithms, polynomial approximation, Fourier approximation, exponential approximation, other topics. Revised and enlarged 2nd edition. 721pp. 5⅜ × 8½.
65241-6 Pa. $14.95

THEORETICAL SOLID STATE PHYSICS, Vol. I: Perfect Lattices in Equilibrium; Vol. II: Non-Equilibrium and Disorder, William Jones and Norman H. March. Monumental reference work covers fundamental theory of equilibrium properties of perfect crystalline solids, non-equilibrium properties, defects and disordered systems. Appendices. Problems. Preface. Diagrams. Index. Bibliography. Total of 1,301pp. 5⅜ × 8½. Two volumes. Vol. I 65015-4 Pa. $14.95
Vol. II 65016-2 Pa. $14.95

OPTIMIZATION THEORY WITH APPLICATIONS, Donald A. Pierre. Broad-spectrum approach to important topic. Classical theory of minima and maxima, calculus of variations, simplex technique and linear programming, more. Many problems, examples. 640pp. 5⅜ × 8½. 65205-X Pa. $14.95

THE MODERN THEORY OF SOLIDS, Frederick Seitz. First inexpensive edition of classic work on theory of ionic crystals, free-electron theory of metals and semiconductors, molecular binding, much more. 736pp. 5⅜ × 8½.
65482-6 Pa. $15.95

ESSAYS ON THE THEORY OF NUMBERS, Richard Dedekind. Two classic essays by great German mathematician: on the theory of irrational numbers; and on transfinite numbers and properties of natural numbers. 115pp. 5⅜ × 8½.
21010-3 Pa. $4.95

THE FUNCTIONS OF MATHEMATICAL PHYSICS, Harry Hochstadt. Comprehensive treatment of orthogonal polynomials, hypergeometric functions, Hill's equation, much more. Bibliography. Index. 322pp. 5⅜ × 8½. 65214-9 Pa. $9.95

NUMBER THEORY AND ITS HISTORY, Oystein Ore. Unusually clear, accessible introduction covers counting, properties of numbers, prime numbers, much more. Bibliography. 380pp. 5⅜ × 8½. 65620-9 Pa. $9.95

THE VARIATIONAL PRINCIPLES OF MECHANICS, Cornelius Lanczos. Graduate level coverage of calculus of variations, equations of motion, relativistic mechanics, more. First inexpensive paperbound edition of classic treatise. Index. Bibliography. 418pp. 5⅜ × 8½. 65067-7 Pa. $11.95

MATHEMATICAL TABLES AND FORMULAS, Robert D. Carmichael and Edwin R. Smith. Logarithms, sines, tangents, trig functions, powers, roots, reciprocals, exponential and hyperbolic functions, formulas and theorems. 269pp. 5⅜ × 8½. 60111-0 Pa. $6.95

THEORETICAL PHYSICS, Georg Joos, with Ira M. Freeman. Classic overview covers essential math, mechanics, electromagnetic theory, thermodynamics, quantum mechanics, nuclear physics, other topics. First paperback edition. xxiii + 885pp. 5⅜ × 8½. 65227-0 Pa. $19.95

HANDBOOK OF MATHEMATICAL FUNCTIONS WITH FORMULAS, GRAPHS, AND MATHEMATICAL TABLES, edited by Milton Abramowitz and Irene A. Stegun. Vast compendium: 29 sets of tables, some to as high as 20 places. 1,046pp. 8 × 10½. 61272-4 Pa. $24.95

MATHEMATICAL METHODS IN PHYSICS AND ENGINEERING, John W. Dettman. Algebraically based approach to vectors, mapping, diffraction, other topics in applied math. Also generalized functions, analytic function theory, more. Exercises. 448pp. 5⅜ × 8¼. 65649-7 Pa. $9.95

A SURVEY OF NUMERICAL MATHEMATICS, David M. Young and Robert Todd Gregory. Broad self-contained coverage of computer-oriented numerical algorithms for solving various types of mathematical problems in linear algebra, ordinary and partial, differential equations, much more. Exercises. Total of 1,248pp. 5⅜ × 8½. Two volumes. Vol. I 65691-8 Pa. $14.95
Vol. II 65692-6 Pa. $14.95

TENSOR ANALYSIS FOR PHYSICISTS, J.A. Schouten. Concise exposition of the mathematical basis of tensor analysis, integrated with well-chosen physical examples of the theory. Exercises. Index. Bibliography. 289pp. 5⅜ × 8½. 65582-2 Pa. $8.95

INTRODUCTION TO NUMERICAL ANALYSIS (2nd Edition), F.B. Hildebrand. Classic, fundamental treatment covers computation, approximation, interpolation, numerical differentiation and integration, other topics. 150 new problems. 669pp. 5⅜ × 8½. 65363-3 Pa. $14.95

INVESTIGATIONS ON THE THEORY OF THE BROWNIAN MOVEMENT, Albert Einstein. Five papers (1905–8) investigating dynamics of Brownian motion and evolving elementary theory. Notes by R. Fürth. 122pp. 5⅜ × 8½. 60304-0 Pa. $4.95

CATASTROPHE THEORY FOR SCIENTISTS AND ENGINEERS, Robert Gilmore. Advanced-level treatment describes mathematics of theory grounded in the work of Poincaré, R. Thom, other mathematicians. Also important applications to problems in mathematics, physics, chemistry and engineering. 1981 edition. References. 28 tables. 397 black-and-white illustrations. xvii + 666pp. 6⅛ × 9¼. 67539-4 Pa. $16.95

AN INTRODUCTION TO STATISTICAL THERMODYNAMICS, Terrell L. Hill. Excellent basic text offers wide-ranging coverage of quantum statistical mechanics, systems of interacting molecules, quantum statistics, more. 523pp. 5⅜ × 8½. 65242-4 Pa. $12.95

ELEMENTARY DIFFERENTIAL EQUATIONS, William Ted Martin and Eric Reissner. Exceptionally clear, comprehensive introduction at undergraduate level. Nature and origin of differential equations, differential equations of first, second and higher orders. Picard's Theorem, much more. Problems with solutions. 331pp. 5⅜ × 8½. 65024-3 Pa. $8.95

STATISTICAL PHYSICS, Gregory H. Wannier. Classic text combines thermodynamics, statistical mechanics and kinetic theory in one unified presentation of thermal physics. Problems with solutions. Bibliography. 532pp. 5⅜ × 8½. 65401-X Pa. $11.95

ORDINARY DIFFERENTIAL EQUATIONS, Morris Tenenbaum and Harry Pollard. Exhaustive survey of ordinary differential equations for undergraduates in mathematics, engineering, science. Thorough analysis of theorems. Diagrams. Bibliography. Index. 818pp. 5⅜ × 8½. 64940-7 Pa. $16.95

STATISTICAL MECHANICS: Principles and Applications, Terrell L. Hill. Standard text covers fundamentals of statistical mechanics, applications to fluctuation theory, imperfect gases, distribution functions, more. 448pp. 5⅜ × 8½. 65390-0 Pa. $9.95

ORDINARY DIFFERENTIAL EQUATIONS AND STABILITY THEORY: An Introduction, David A. Sánchez. Brief, modern treatment. Linear equation, stability theory for autonomous and nonautonomous systems, etc. 164pp. 5⅜ × 8¼. 63828-6 Pa. $5.95

THIRTY YEARS THAT SHOOK PHYSICS: The Story of Quantum Theory, George Gamow. Lucid, accessible introduction to influential theory of energy and matter. Careful explanations of Dirac's anti-particles, Bohr's model of the atom, much more. 12 plates. Numerous drawings. 240pp. 5⅜ × 8½. 24895-X Pa. $6.95

THEORY OF MATRICES, Sam Perlis. Outstanding text covering rank, non-singularity and inverses in connection with the development of canonical matrices under the relation of equivalence, and without the intervention of determinants. Includes exercises. 237pp. 5⅜ × 8½. 66810-X Pa. $7.95

GREAT EXPERIMENTS IN PHYSICS: Firsthand Accounts from Galileo to Einstein, edited by Morris H. Shamos. 25 crucial discoveries: Newton's laws of motion, Chadwick's study of the neutron, Hertz on electromagnetic waves, more. Original accounts clearly annotated. 370pp. 5⅜ × 8½. 25346-5 Pa. $10.95

INTRODUCTION TO PARTIAL DIFFERENTIAL EQUATIONS WITH APPLICATIONS, E.C. Zachmanoglou and Dale W. Thoe. Essentials of partial differential equations applied to common problems in engineering and the physical sciences. Problems and answers. 416pp. 5⅜ × 8½. 65251-3 Pa. $10.95

BURNHAM'S CELESTIAL HANDBOOK, Robert Burnham, Jr. Thorough guide to the stars beyond our solar system. Exhaustive treatment. Alphabetical by constellation: Andromeda to Cetus in Vol. 1; Chamaeleon to Orion in Vol. 2; and Pavo to Vulpecula in Vol. 3. Hundreds of illustrations. Index in Vol. 3. 2,000pp. 6⅛ × 9¼. 23567-X, 23568-8, 23673-0 Pa., Three-vol. set $41.85

CHEMICAL MAGIC, Leonard A. Ford. Second Edition, Revised by E. Winston Grundmeier. Over 100 unusual stunts demonstrating cold fire, dust explosions, much more. Text explains scientific principles and stresses safety precautions. 128pp. 5⅜ × 8½. 67628-5 Pa. $5.95

AMATEUR ASTRONOMER'S HANDBOOK, J.B. Sidgwick. Timeless, comprehensive coverage of telescopes, mirrors, lenses, mountings, telescope drives, micrometers, spectroscopes, more. 189 illustrations. 576pp. 5⅜ × 8¼. (Available in U.S. only) 24034-7 Pa. $9.95

SPECIAL FUNCTIONS, N.N. Lebedev. Translated by Richard Silverman. Famous Russian work treating more important special functions, with applications to specific problems of physics and engineering. 38 figures. 308pp. 5⅜ × 8½.
60624-4 Pa. $8.95

OBSERVATIONAL ASTRONOMY FOR AMATEURS, J.B. Sidgwick. Mine of useful data for observation of sun, moon, planets, asteroids, aurorae, meteors, comets, variables, binaries, etc. 39 illustrations. 384pp. 5⅜ × 8¼. (Available in U.S. only)
24033-9 Pa. $8.95

INTEGRAL EQUATIONS, F.G. Tricomi. Authoritative, well-written treatment of extremely useful mathematical tool with wide applications. Volterra Equations, Fredholm Equations, much more. Advanced undergraduate to graduate level. Exercises. Bibliography. 238pp. 5⅜ × 8½.
64828-1 Pa. $7.95

POPULAR LECTURES ON MATHEMATICAL LOGIC, Hao Wang. Noted logician's lucid treatment of historical developments, set theory, model theory, recursion theory and constructivism, proof theory, more. 3 appendixes. Bibliography. 1981 edition. ix + 283pp. 5⅜ × 8½.
67632-3 Pa. $8.95

MODERN NONLINEAR EQUATIONS, Thomas L. Saaty. Emphasizes practical solution of problems; covers seven types of equations. ". . . a welcome contribution to the existing literature. . . ."—*Math Reviews.* 490pp. 5⅜ × 8½. 64232-1 Pa. $11.95

FUNDAMENTALS OF ASTRODYNAMICS, Roger Bate et al. Modern approach developed by U.S. Air Force Academy. Designed as a first course. Problems, exercises. Numerous illustrations. 455pp. 5⅜ × 8½.
60061-0 Pa. $9.95

INTRODUCTION TO LINEAR ALGEBRA AND DIFFERENTIAL EQUATIONS, John W. Dettman. Excellent text covers complex numbers, determinants, orthonormal bases, Laplace transforms, much more. Exercises with solutions. Undergraduate level. 416pp. 5⅜ × 8½.
65191-6 Pa. $9.95

INCOMPRESSIBLE AERODYNAMICS, edited by Bryan Thwaites. Covers theoretical and experimental treatment of the uniform flow of air and viscous fluids past two-dimensional aerofoils and three-dimensional wings; many other topics. 654pp. 5⅜ × 8½.
65465-6 Pa. $16.95

INTRODUCTION TO DIFFERENCE EQUATIONS, Samuel Goldberg. Exceptionally clear exposition of important discipline with applications to sociology, psychology, economics. Many illustrative examples; over 250 problems. 260pp. 5⅜ × 8½.
65084-7 Pa. $7.95

LAMINAR BOUNDARY LAYERS, edited by L. Rosenhead. Engineering classic covers steady boundary layers in two- and three-dimensional flow, unsteady boundary layers, stability, observational techniques, much more. 708pp. 5⅜ × 8½.
65646-2 Pa. $18.95

LECTURES ON CLASSICAL DIFFERENTIAL GEOMETRY, Second Edition, Dirk J. Struik. Excellent brief introduction covers curves, theory of surfaces, fundamental equations, geometry on a surface, conformal mapping, other topics. Problems. 240pp. 5⅜ × 8½.
65609-8 Pa. $7.95

ROTARY-WING AERODYNAMICS, W.Z. Stepniewski. Clear, concise text covers aerodynamic phenomena of the rotor and offers guidelines for helicopter performance evaluation. Originally prepared for NASA. 537 figures. 640pp. 6⅛ × 9¼.
64647-5 Pa. $15.95

DIFFERENTIAL GEOMETRY, Heinrich W. Guggenheimer. Local differential geometry as an application of advanced calculus and linear algebra. Curvature, transformation groups, surfaces, more. Exercises. 62 figures. 378pp. 5⅜ × 8½.
63433-7 Pa. $8.95

INTRODUCTION TO SPACE DYNAMICS, William Tyrrell Thomson. Comprehensive, classic introduction to space-flight engineering for advanced undergraduate and graduate students. Includes vector algebra, kinematics, transformation of coordinates. Bibliography. Index. 352pp. 5⅜ × 8½. 65113-4 Pa. $8.95

A SURVEY OF MINIMAL SURFACES, Robert Osserman. Up-to-date, in-depth discussion of the field for advanced students. Corrected and enlarged edition covers new developments. Includes numerous problems. 192pp. 5⅜ × 8½.
64998-9 Pa. $8.95

ANALYTICAL MECHANICS OF GEARS, Earle Buckingham. Indispensable reference for modern gear manufacture covers conjugate gear-tooth action, gear-tooth profiles of various gears, many other topics. 263 figures. 102 tables. 546pp. 5⅜ × 8½. 65712-4 Pa. $14.95

SET THEORY AND LOGIC, Robert R. Stoll. Lucid introduction to unified theory of mathematical concepts. Set theory and logic seen as tools for conceptual understanding of real number system. 496pp. 5⅜ × 8¼. 63829-4 Pa. $10.95

A HISTORY OF MECHANICS, René Dugas. Monumental study of mechanical principles from antiquity to quantum mechanics. Contributions of ancient Greeks, Galileo, Leonardo, Kepler, Lagrange, many others. 671pp. 5⅜ × 8½.
65632-2 Pa. $14.95

FAMOUS PROBLEMS OF GEOMETRY AND HOW TO SOLVE THEM, Benjamin Bold. Squaring the circle, trisecting the angle, duplicating the cube: learn their history, why they are impossible to solve, then solve them yourself. 128pp. 5⅜ × 8½. 24297-8 Pa. $4.95

MECHANICAL VIBRATIONS, J.P. Den Hartog. Classic textbook offers lucid explanations and illustrative models, applying theories of vibrations to a variety of practical industrial engineering problems. Numerous figures. 233 problems, solutions. Appendix. Index. Preface. 436pp. 5⅜ × 8½. 64785-4 Pa. $10.95

CURVATURE AND HOMOLOGY, Samuel I. Goldberg. Thorough treatment of specialized branch of differential geometry. Covers Riemannian manifolds, topology of differentiable manifolds, compact Lie groups, other topics. Exercises. 315pp. 5⅜ × 8½. 64314-X Pa. $8.95

HISTORY OF STRENGTH OF MATERIALS, Stephen P. Timoshenko. Excellent historical survey of the strength of materials with many references to the theories of elasticity and structure. 245 figures. 452pp. 5⅜ × 8½. 61187-6 Pa. $11.95

CATALOG OF DOVER BOOKS

GEOMETRY OF COMPLEX NUMBERS, Hans Schwerdtfeger. Illuminating, widely praised book on analytic geometry of circles, the Moebius transformation, and two-dimensional non-Euclidean geometries. 200pp. 5⅜ × 8¼.

63830-8 Pa. $8.95

MECHANICS, J.P. Den Hartog. A classic introductory text or refresher. Hundreds of applications and design problems illuminate fundamentals of trusses, loaded beams and cables, etc. 334 answered problems. 462pp. 5⅜ × 8½.   60754-2 Pa. $9.95

TOPOLOGY, John G. Hocking and Gail S. Young. Superb one-year course in classical topology. Topological spaces and functions, point-set topology, much more. Examples and problems. Bibliography. Index. 384pp. 5⅜ × 8¼.

65676-4 Pa. $9.95

STRENGTH OF MATERIALS, J.P. Den Hartog. Full, clear treatment of basic material (tension, torsion, bending, etc.) plus advanced material on engineering methods, applications. 350 answered problems. 323pp. 5⅜ × 8½. 60755-0 Pa. $8.95

ELEMENTARY CONCEPTS OF TOPOLOGY, Paul Alexandroff. Elegant, intuitive approach to topology from set-theoretic topology to Betti groups; how concepts of topology are useful in math and physics. 25 figures. 57pp. 5⅜ × 8½.

60747-X Pa. $3.50

ADVANCED STRENGTH OF MATERIALS, J.P. Den Hartog. Superbly written advanced text covers torsion, rotating disks, membrane stresses in shells, much more. Many problems and answers. 388pp. 5⅜ × 8½.   65407-9 Pa. $9.95

COMPUTABILITY AND UNSOLVABILITY, Martin Davis. Classic graduate-level introduction to theory of computability, usually referred to as theory of recurrent functions. New preface and appendix. 288pp. 5⅜ × 8½. 61471-9 Pa. $7.95

GENERAL CHEMISTRY, Linus Pauling. Revised 3rd edition of classic first-year text by Nobel laureate. Atomic and molecular structure, quantum mechanics, statistical mechanics, thermodynamics correlated with descriptive chemistry. Problems. 992pp. 5⅜ × 8½.   65622-5 Pa. $19.95

AN INTRODUCTION TO MATRICES, SETS AND GROUPS FOR SCIENCE STUDENTS, G. Stephenson. Concise, readable text introduces sets, groups, and most importantly, matrices to undergraduate students of physics, chemistry, and engineering. Problems. 164pp. 5⅜ × 8½.   65077-4 Pa. $6.95

THE HISTORICAL BACKGROUND OF CHEMISTRY, Henry M. Leicester. Evolution of ideas, not individual biography. Concentrates on formulation of a coherent set of chemical laws. 260pp. 5⅜ × 8½.   61053-5 Pa. $6.95

THE PHILOSOPHY OF MATHEMATICS: An Introductory Essay, Stephan Körner. Surveys the views of Plato, Aristotle, Leibniz & Kant concerning propositions and theories of applied and pure mathematics. Introduction. Two appendices. Index. 198pp. 5⅜ × 8½.   25048-2 Pa. $7.95

THE DEVELOPMENT OF MODERN CHEMISTRY, Aaron J. Ihde. Authoritative history of chemistry from ancient Greek theory to 20th-century innovation. Covers major chemists and their discoveries. 209 illustrations. 14 tables. Bibliographies. Indices. Appendices. 851pp. 5⅜ × 8½.   64235-6 Pa. $18.95

DE RE METALLICA, Georgius Agricola. The famous Hoover translation of greatest treatise on technological chemistry, engineering, geology, mining of early modern times (1556). All 289 original woodcuts. 638pp. 6¾ × 11.
60006-8 Pa. $18.95

SOME THEORY OF SAMPLING, William Edwards Deming. Analysis of the problems, theory and design of sampling techniques for social scientists, industrial managers and others who find statistics increasingly important in their work. 61 tables. 90 figures. xvii + 602pp. 5⅜ × 8½. 64684-X Pa. $15.95

THE VARIOUS AND INGENIOUS MACHINES OF AGOSTINO RAMELLI: A Classic Sixteenth-Century Illustrated Treatise on Technology, Agostino Ramelli. One of the most widely known and copied works on machinery in the 16th century. 194 detailed plates of water pumps, grain mills, cranes, more. 608pp. 9 × 12.
25497-6 Clothbd. $34.95

LINEAR PROGRAMMING AND ECONOMIC ANALYSIS, Robert Dorfman, Paul A. Samuelson and Robert M. Solow. First comprehensive treatment of linear programming in standard economic analysis. Game theory, modern welfare economics, Leontief input-output, more. 525pp. 5⅜ × 8½. 65491-5 Pa. $14.95

ELEMENTARY DECISION THEORY, Herman Chernoff and Lincoln E. Moses. Clear introduction to statistics and statistical theory covers data processing, probability and random variables, testing hypotheses, much more. Exercises. 364pp. 5⅜ × 8½. 65218-1 Pa. $9.95

THE COMPLEAT STRATEGYST: Being a Primer on the Theory of Games of Strategy, J.D. Williams. Highly entertaining classic describes, with many illustrated examples, how to select best strategies in conflict situations. Prefaces. Appendices. 268pp. 5⅜ × 8½. 25101-2 Pa. $7.95

MATHEMATICAL METHODS OF OPERATIONS RESEARCH, Thomas L. Saaty. Classic graduate-level text covers historical background, classical methods of forming models, optimization, game theory, probability, queueing theory, much more. Exercises. Bibliography. 448pp. 5⅜ × 8¼. 65703-5 Pa. $12.95

CONSTRUCTIONS AND COMBINATORIAL PROBLEMS IN DESIGN OF EXPERIMENTS, Damaraju Raghavarao. In-depth reference work examines orthogonal Latin squares, incomplete block designs, tactical configuration, partial geometry, much more. Abundant explanations, examples. 416pp. 5⅜ × 8¼.
65685-3 Pa. $10.95

THE ABSOLUTE DIFFERENTIAL CALCULUS (CALCULUS OF TENSORS), Tullio Levi-Civita. Great 20th-century mathematician's classic work on material necessary for mathematical grasp of theory of relativity. 452pp. 5⅜ × 8½.
63401-9 Pa. $9.95

VECTOR AND TENSOR ANALYSIS WITH APPLICATIONS, A.I. Borisenko and I.E. Tarapov. Concise introduction. Worked-out problems, solutions, exercises. 257pp. 5⅜ × 8¼. 63833-2 Pa. $7.95

THE FOUR-COLOR PROBLEM: Assaults and Conquest, Thomas L. Saaty and Paul G. Kainen. Engrossing, comprehensive account of the century-old combinatorial topological problem, its history and solution. Bibliographies. Index. 110 figures. 228pp. 5⅜ × 8½. 65092-8 Pa. $6.95

CATALYSIS IN CHEMISTRY AND ENZYMOLOGY, William P. Jencks. Exceptionally clear coverage of mechanisms for catalysis, forces in aqueous solution, carbonyl- and acyl-group reactions, practical kinetics, more. 864pp. 5⅜ × 8½. 65460-5 Pa. $19.95

PROBABILITY: An Introduction, Samuel Goldberg. Excellent basic text covers set theory, probability theory for finite sample spaces, binomial theorem, much more. 360 problems. Bibliographies. 322pp. 5⅜ × 8½. 65252-1 Pa. $8.95

LIGHTNING, Martin A. Uman. Revised, updated edition of classic work on the physics of lightning. Phenomena, terminology, measurement, photography, spectroscopy, thunder, more. Reviews recent research. Bibliography. Indices. 320pp. 5⅜ × 8¼. 64575-4 Pa. $8.95

PROBABILITY THEORY: A Concise Course, Y.A. Rozanov. Highly readable, self-contained introduction covers combination of events, dependent events, Bernoulli trials, etc. Translation by Richard Silverman. 148pp. 5⅜ × 8¼. 63544-9 Pa. $5.95

AN INTRODUCTION TO HAMILTONIAN OPTICS, H. A. Buchdahl. Detailed account of the Hamiltonian treatment of aberration theory in geometrical optics. Many classes of optical systems defined in terms of the symmetries they possess. Problems with detailed solutions. 1970 edition. xv + 360pp. 5⅜ × 8½. 67597-1 Pa. $10.95

STATISTICS MANUAL, Edwin L. Crow, et al. Comprehensive, practical collection of classical and modern methods prepared by U.S. Naval Ordnance Test Station. Stress on use. Basics of statistics assumed. 288pp. 5⅜ × 8½. 60599-X Pa. $6.95

DICTIONARY/OUTLINE OF BASIC STATISTICS, John E. Freund and Frank J. Williams. A clear concise dictionary of over 1,000 statistical terms and an outline of statistical formulas covering probability, nonparametric tests, much more. 208pp. 5⅜ × 8½. 66796-0 Pa. $6.95

STATISTICAL METHOD FROM THE VIEWPOINT OF QUALITY CONTROL, Walter A. Shewhart. Important text explains regulation of variables, uses of statistical control to achieve quality control in industry, agriculture, other areas. 192pp. 5⅜ × 8½. 65232-7 Pa. $7.95

THE INTERPRETATION OF GEOLOGICAL PHASE DIAGRAMS, Ernest G. Ehlers. Clear, concise text emphasizes diagrams of systems under fluid or containing pressure; also coverage of complex binary systems, hydrothermal melting, more. 288pp. 6½ × 9¼. 65389-7 Pa. $10.95

STATISTICAL ADJUSTMENT OF DATA, W. Edwards Deming. Introduction to basic concepts of statistics, curve fitting, least squares solution, conditions without parameter, conditions containing parameters. 26 exercises worked out. 271pp. 5⅜ × 8½. 64685-8 Pa. $8.95

TENSOR CALCULUS, J.L. Synge and A. Schild. Widely used introductory text covers spaces and tensors, basic operations in Riemannian space, non-Riemannian spaces, etc. 324pp. 5⅜ × 8¼. 63612-7 Pa. $8.95

A CONCISE HISTORY OF MATHEMATICS, Dirk J. Struik. The best brief history of mathematics. Stresses origins and covers every major figure from ancient Near East to 19th century. 41 illustrations. 195pp. 5⅜ × 8½. 60255-9 Pa. $7.95

A SHORT ACCOUNT OF THE HISTORY OF MATHEMATICS, W.W. Rouse Ball. One of clearest, most authoritative surveys from the Egyptians and Phoenicians through 19th-century figures such as Grassman, Galois, Riemann. Fourth edition. 522pp. 5⅜ × 8½. 20630-0 Pa. $10.95

HISTORY OF MATHEMATICS, David E. Smith. Nontechnical survey from ancient Greece and Orient to late 19th century; evolution of arithmetic, geometry, trigonometry, calculating devices, algebra, the calculus. 362 illustrations. 1,355pp. 5⅜ × 8½. 20429-4, 20430-8 Pa., Two-vol. set $23.90

THE GEOMETRY OF RENÉ DESCARTES, René Descartes. The great work founded analytical geometry. Original French text, Descartes' own diagrams, together with definitive Smith-Latham translation. 244pp. 5⅜ × 8½.
60068-8 Pa. $6.95

THE ORIGINS OF THE INFINITESIMAL CALCULUS, Margaret E. Baron. Only fully detailed and documented account of crucial discipline: origins; development by Galileo, Kepler, Cavalieri; contributions of Newton, Leibniz, more. 304pp. 5⅜ × 8½. (Available in U.S. and Canada only) 65371-4 Pa. $9.95

THE HISTORY OF THE CALCULUS AND ITS CONCEPTUAL DEVELOPMENT, Carl B. Boyer. Origins in antiquity, medieval contributions, work of Newton, Leibniz, rigorous formulation. Treatment is verbal. 346pp. 5⅜ × 8½.
60509-4 Pa. $8.95

THE THIRTEEN BOOKS OF EUCLID'S ELEMENTS, translated with introduction and commentary by Sir Thomas L. Heath. Definitive edition. Textual and linguistic notes, mathematical analysis. 2,500 years of critical commentary. Not abridged. 1,414pp. 5⅜ × 8½. 60088-2, 60089-0, 60090-4 Pa., Three-vol. set $29.85

GAMES AND DECISIONS: Introduction and Critical Survey, R. Duncan Luce and Howard Raiffa. Superb nontechnical introduction to game theory, primarily applied to social sciences. Utility theory, zero-sum games, n-person games, decision-making, much more. Bibliography. 509pp. 5⅜ × 8½. 65943-7 Pa. $12.95

THE HISTORICAL ROOTS OF ELEMENTARY MATHEMATICS, Lucas N.H. Bunt, Phillip S. Jones, and Jack D. Bedient. Fundamental underpinnings of modern arithmetic, algebra, geometry and number systems derived from ancient civilizations. 320pp. 5⅜ × 8½. 25563-8 Pa. $8.95

CALCULUS REFRESHER FOR TECHNICAL PEOPLE, A. Albert Klaf. Covers important aspects of integral and differential calculus via 756 questions. 566 problems, most answered. 431pp. 5⅜ × 8½. 20370-0 Pa. $8.95

CHALLENGING MATHEMATICAL PROBLEMS WITH ELEMENTARY SOLUTIONS, A.M. Yaglom and I.M. Yaglom. Over 170 challenging problems on probability theory, combinatorial analysis, points and lines, topology, convex polygons, many other topics. Solutions. Total of 445pp. 5⅜ × 8½. Two-vol. set.
Vol. I 65536-9 Pa. $7.95
Vol. II 65537-7 Pa. $6.95

FIFTY CHALLENGING PROBLEMS IN PROBABILITY WITH SOLUTIONS, Frederick Mosteller. Remarkable puzzlers, graded in difficulty, illustrate elementary and advanced aspects of probability. Detailed solutions. 88pp. 5⅜ × 8½.
65355-2 Pa. $4.95

EXPERIMENTS IN TOPOLOGY, Stephen Barr. Classic, lively explanation of one of the byways of mathematics. Klein bottles, Moebius strips, projective planes, map coloring, problem of the Koenigsberg bridges, much more, described with clarity and wit. 43 figures. 210pp. 5⅜ × 8½. 25933-1 Pa. $5.95

RELATIVITY IN ILLUSTRATIONS, Jacob T. Schwartz. Clear nontechnical treatment makes relativity more accessible than ever before. Over 60 drawings illustrate concepts more clearly than text alone. Only high school geometry needed. Bibliography. 128pp. 6⅛ × 9¼. 25965-X Pa. $6.95

AN INTRODUCTION TO ORDINARY DIFFERENTIAL EQUATIONS, Earl A. Coddington. A thorough and systematic first course in elementary differential equations for undergraduates in mathematics and science, with many exercises and problems (with answers). Index. 304pp. 5⅜ × 8½. 65942-9 Pa. $8.95

FOURIER SERIES AND ORTHOGONAL FUNCTIONS, Harry F. Davis. An incisive text combining theory and practical example to introduce Fourier series, orthogonal functions and applications of the Fourier method to boundary-value problems. 570 exercises. Answers and notes. 416pp. 5⅜ × 8½. 65973-9 Pa. $9.95

THE THEORY OF BRANCHING PROCESSES, Theodore E. Harris. First systematic, comprehensive treatment of branching (i.e. multiplicative) processes and their applications. Galton-Watson model, Markov branching processes, electron-photon cascade, many other topics. Rigorous proofs. Bibliography. 240pp. 5⅜ × 8½. 65952-6 Pa. $6.95

AN INTRODUCTION TO ALGEBRAIC STRUCTURES, Joseph Landin. Superb self-contained text covers "abstract algebra": sets and numbers, theory of groups, theory of rings, much more. Numerous well-chosen examples, exercises. 247pp. 5⅜ × 8½. 65940-2 Pa. $7.95
